The Rise of
the Corporate Economy

ນລ

The Rise of
the Corporate Economy

The British Experience

LESLIE HANNAH

೫೫

Foreword by
Alfred D. Chandler, Jr

೫೫

The Johns Hopkins University Press
Baltimore and London

To my mother
and to the memory of
my father

ನಲ್

Contents

ౠ

න්‍

The fundamental problem ... of the
social science, is to find the laws according
to which any state of society produces the
state which succeeds it and takes its place.
This opens the great and vexed question
of the progressiveness of man and
society; an idea involved in every just
conception of social phenomena as the
subject of science.

J. S. MILL, *System of Logic* (1852 edition) p. 503

න්‍

Foreword

ℜℜ

In both method and substance Leslie Hannah's *Rise of the Corporate Economy* has much to offer American readers. Hannah has combined accurate historical narrative with effective economic analysis to produce a model of what can be called institutional economic history. The resulting text permits for the first time a comparison of institutional change in the American economy with that of Britain, the world's first industrial nation.

The similarities between the British and American experiences are striking and the differences revealing. In both countries the large corporate enterprise played a fundamental role in the modernization of the economies. In the nineteenth century such firms came to dominate the new railroad transportation. In the twentieth century they began to do the same in production and distribution. By 1930 in both economies the 100 largest enterprises accounted for about 25 per cent of the net output in the manufacturing sector and by 1960 for well over 30 per cent. In both countries large corporate enterprises came to cluster in much the same technologically advanced industries – industries that were central to the continuing health and growth of an industrial economy. These industries – chemicals, oil, rubber, metal making, automobiles, electrical and electronic machinery, and some foods – were from their earliest years highly concentrated. Their structure, however, became more oligopolistic than monopolistic. In both economies the first firms to grow large through the exploitation of a new technology remained the leading firms in their industries. Such 'core' enterprises continued to grow through internal growth and through merger. Finally in both countries the merger movements appeared almost simultaneously – at the turn of the century, in the 1920s, and again in the 1960s. These similarities suggest what may be common patterns of institutional change in the growth of technologically advanced industrial, urban economies.

Nevertheless, the differences are significant. In Britain more firms grew to large size through merger than they did in the United States. Administrative centralization and vertical integration came much more slowly than they did in American mergers. Large British holding companies often remained little more than federations of family firms. In Britain owners continued to manage much longer than they did in the United States. By mid-century when the United States had become the home of managerial capitalism, Britain remained a bastion of family capitalism.

Thus, if the similarities indicate the organizational imperatives of a modern economy, the differences suggest the ways in which resource endowment, markets, social and class structure, and cultural attitudes and values affect the institutional arrangements through which such economies are managed. The challenge posed by Hannah's book should stimulate Americans to undertake similar careful and detailed studies of institutional change in their own economy. Such analysis, by sharpening and making more precise likenesses and contrasts, can contribute enormously to the understanding of the recent economic past of both nations.

Alfred D. Chandler, Jr

Acknowledgements

ಬಿ

I have been fortunate in accumulating a large overdraft of intellectual debts to tolerant fellow economists and historians, particularly to past and present colleagues at Nuffield College and St John's College, Oxford, and at the University of Essex. Max Hartwell, William Reader, George Richardson, Aubrey Silberston and John Wright have at various times stimulated my interest in the subject, and their critical reading of earlier drafts of some chapters led to improvements. Many others have helped with specific problems, including Sam Aaronovitch, G. C. Allen, Alfred Chandler, Denys Gribbin, Peter Hammond, Lutz Haber, Peter Hart, William Kennedy, Jeremy Lever, David Lethbridge, Peter Mathias, John Naylor, Ralph Nelson, Peter Payne, Sigbert Prais, Hilary Rubinstein, Clive Trebilcock and Philip Williams. My especial thanks are due to John Kay, who read and helpfully criticized earlier drafts and also permitted the use of previously unpublished results from our joint work on mergers and industrial concentration. Many industrialists and politicians who have taken part in the events analysed have also given me their time and advice, and I am particularly grateful to Arthur Knight, the late Lord Swinton, Lyndall Urwick, Harry Ward, the late Sir Horace Wilson and Arthur Young, and to the companies and individuals who gave me access to private collections of papers not normally available for public perusal. Much of the statistical work incorporated in the appendices was financed by the Nuffield Foundation and the Social Science Research Council, and skilfully executed by Margaret Ackrill and Patricia Wright, who also provided constructive suggestions. The editors of *Business History*, the *Economic History Review* and *Oxford Economic Papers* kindly permitted me to reprint passages from articles originally published in their journals. A difficult manuscript became a readable typescript through the capable ministrations of Alison Hunt, Beryce Vincenzi and Sue Dawson.

I am grateful to them all, but these acknowledgements do not, of course, implicate them in any errors or omissions which remain. The omissions, at least, may be cheerfully admitted, for our knowledge of many aspects of the rise of the corporate economy will surely be further enlightened, not least by the ongoing work of those acknowledged here.

Wivenhoe Park
Colchester

I

Business: history and economics

Just as microscopic work on cells may throw new light on
the human body, so detailed study of the growth of
particular business units may add to knowledge
of the industrial system.

T. S. ASHTON, *An Eighteenth Century Industrialist:
Peter Stubs of Warrington* (Manchester, 1939), p. ix.

ೞ

It is a commonplace that in the course of the present century British
industry has witnessed a transformation from a disaggregated structure
of predominantly small, competing firms to a concentrated structure
dominated by large, and often monopolistic, corporations. The 100
corporations which now occupy the dominant position account for
something approaching one half of total manufacturing output,[1] whilst
at the turn of the century the largest 100 firms accounted for barely 15
per cent of output (see Appendix 2). The large enterprises of today
have achieved this increase in their relative economic power partly by
internal expansion and partly by the absorption and competitive
elimination of smaller rivals. The number of small firms, employing
200 people or less, which was already only 136,000 in 1935 had by
1963 declined to 60,000, and their share in manufacturing output over
the same period was reduced from 35 to only 16 per cent.[2] In qualitative terms also the firms of today, and particularly the larger ones,
differ significantly from their Victorian forebears: they are more
diversified, they have more complex organization structures, they spend
more on research, they are more likely to acquire control of rivals, and
they are now themselves more frequently the subject of takeover bids.
Many of their products are also very different from those of Victorian
firms, for the development of large corporations is closely bound up
with the 'second industrial revolution', with twentieth century economic
growth based on electricity, the motor car and chemicals, rather than

[1] This book mainly confines itself to trends in manufacturing industry, though
similar trends have occurred in most other sectors of the economy.
[2] (Bolton) Committee of Inquiry on Small Firms, *Report* (Cmd. 4811, 1971),
pp. 59–60.

steam, railways and textiles. Precisely when, and in what stages, such developments took place, and the underlying reasons for the transformation in economic life which they implied, will be a major theme in the chapters that follow.

These developments were not, of course, confined to the United Kingdom; indeed, the tendency to increasing industrial concentration is one of the better attested facts of the recent economic history of most economically advanced Western countries. Moreover, in Britain as in America, the large corporations themselves have been no respecters of national boundaries, but have expanded overseas as well as in the domestic economy. By the 1970s the sales of the largest corporations in the United States and Europe exceeded the national income of many of the member states of the United Nations. The development of these corporations has, therefore, been a major preoccupation of economists, and the importance of an understanding of these changes can scarcely be exaggerated.

The significance of this organizational transformation extends beyond the economic sphere; there are profound political and sociological implications also. It has facilitated, and perhaps induced, substantial changes in the relationship between government and industry. It has brought an increasing number of workers into the employment of large organizations. It has extended to the wealthier middle class what the industrial revolution accomplished for the working class, by breaking the links between family and work, which survived in the Victorian family business but are increasingly rare today. In making this break, it has divorced the role of saving and investing from that of managing and directing, yet it has nonetheless preserved substantial inequalities of wealth and power. Despite the persistence of such inequalities, the economic success of managerial capitalism appears to have reconciled significant numbers of the politically conscious electorates of industrialized nations to a programme of modifying capitalism by political agreement rather than by revolutionary change. The harshnesses of capitalism that remain may still bear down heavily on individuals, but they now do so less as a result of competitive market pressures on employment and wages, and more as a result of decisions which emanate from a managerial hierarchy which has supplemented the market as a means of coordinating economic activities. Authority structures and managerial prerogatives, as well as the market itself, are now commonly (and properly) the focus of moral and social criticism of the economic system.

The economic significance of the rise of the corporate economy is also rooted in the replacement of the 'invisible hand' of the market place by the more conscious integration and organization of economic activities within large firms. In the nineteenth century the flow of production, from the winning of raw materials to the final sale to the consumer, typically involved many middlemen, each with a specialized function; and resources for production were, broadly, allocated by a market mechanism mediating the relationships between many producers and consumers. Increasingly, now, however, resources and intermediate products are allocated not by dealings between firms in markets, but rather within firms – by an administrative mechanism rather than by a price mechanism. Indeed, the rationale of a firm is that there are costs to using the price mechanism – transactions costs – on which a firm can economize, and 'the distinguishing mark of the firm is the supersession of the price mechanism'.[3] The boundaries of a firm are determined by the relative costs of market and firm, as businessmen continually adjust the activity of their firms at the margin in response to these relative costs. Among the private costs faced by a businessman using the market, for example, might be listed uncertainty, contractual costs, advertising and other sales costs. In a broader perspective, a businessman might count among the costs of the market the fact that, by competing with other firms in a market, he has to forgo externalities, economies of scale or monopoly profits, which would be possible if he enlarged his firm. The increased role of large firms in economic activity does, then, suggest that the private benefits of planning and organizing production within the firm have been increasing relative to the advantages of dealing in markets. The head offices of the large corporations, which now rise on the banks of the Thames near the administrative and political heartland of London, perform much of the work of organization and coordination formerly achieved less conspicuously by market transactions. They thus fulfil as important a role in modern economic life as the commodity exchanges and other venues for market transactions of the nineteenth century economy.

The precise implication of this change remains a matter of much dispute. It has, for example, been suggested that the managers who perform the organizational task – Professor Galbraith's 'technostructure'[4] – have objectives differing from those of the classical owner-

[3] R. H. Coase, 'The nature of the firm', *Economica*, vol. 4 (1937), as reprinted in American Economic Association (ed.), *Readings in Price Theory* (1953), p. 334.   [4] J. K. Galbraith, *The New Industrial State* (1967).

entrepreneur, and that it is now they, rather than wealth owners, who
determine major economic choices. On the other hand many markets
still exist, and the discipline on firms in the final product market, not
to speak of the threat of takeover in the capital market, can still set
limits to the freedom of action of managers even in the modern corporate economy. Moreover, there are, in fact, many ways in which
modern corporations can, in their internal decision making, mimic the
rules of the price mechanism which they have replaced. Within a large
corporation, for example, a number of managerially independent
divisions normally compete with each other for resources from head
office. In a way reminiscent of the allocation of resources by the capital
market among firms, the head office will normally assess requests from
divisions according to past performance and likely future profitability,
so that the resulting economic decisions may not differ markedly from
those in the capital market. Both systems may, in fact, be directed
towards maximizing returns on investments and, if this is so, it is the
means rather than the end that has changed. Such themes will naturally
recur in the following historical analysis of the evolution of the corporate economy.

A crucial factor in determining the relative costs of market and firm
is management itself. Fortunately, with the recent publication of a
number of business histories, we can now make a clearer assessment of
the significance of changes in management costs and practices. Although
Professor Charles Wilson, in the preface to his pioneering *History of
Unilever*, modestly averred that 'it is not necessary to claim that business
history is history of the highest calibre',[5] his apology now seems un-
necessary in view of his own contribution and a run of works of comparable quality culminating in recent years in Professor Coleman's
Courtaulds and Dr Reader's *ICI*.[6] Given the distance in time from the
decisions which such histories record, they are usually able to provide
for the historian a franker analysis of events than is available to the
student of contemporary business decisions, and they provide a detailed
and suggestive treatment of the strategy and structure of major enterprises in the corporate economy. Yet business historians, who in general
quite properly prize their allegiance to the facts of the individual case,
have sometimes tended to see events in the history of the firm as unique,

[5] C. Wilson, *History of Unilever*, vol. 1 (1954), p. ix.
[6] D. C. Coleman, *Courtaulds, An Economic and Social History*, 2 vols (Oxford,
1969). W. J. Reader, *Imperial Chemical Industries: A History*, vol. 1 (1970),
vol. 2 (1975).

biographical events explicable in terms of particular or even accidental concatenations of historical forces. They rightly reject econometric reasoning as inappropriate to the single case, but the absence of generality has perhaps unnecessarily inhibited business historians in the rigorous development of causal interpretations.

In Britain,[7] then, it has fallen to an economist, Professor Edith Penrose, to exploit historical case study material and to develop a model of the growth of the firm. In her important contribution to the dynamic theory of the firm, however, she wisely warned her readers of the difficulty of developing effective generalizations from the biased sample of case histories which is available:

> Consistent examples are [she admitted] . . . no more a proof than are inconsistent examples a disproof of a general argument unless the examples are presented in sufficiently large numbers and selected in such a way that they constitute a representative sample . . . the examples presented in the following chapters are illustrative only. There is not sufficient systematic information available as yet to enable any comprehensive testing of the generality of the theory.[8]

This *caveat* also applies to many of the generalizations below. However, by focusing attention on mergers and industrial concentration, it has been possible to assemble fuller statistical data (which are presented in the Appendices) to strengthen the case study analysis of the growth of firms. Moreover, with the recent release of much new archival material by both business and government,[9] the generalizations which have in the past been made about the forces making for increased industrial concentration are now more tractable to systematic analysis. Even so, there remain many dark areas where our knowledge is limited, and inferences drawn from contemporary views, or from incomplete or biased data, are still in some cases the only basis of interpretation which is available.

[7] But not in the United States, where Professor A. D. Chandler has made a classic historical study of the American corporate economy in his *Strategy and Structure, Chapters in the History of Industrial Enterprise* (Cambridge, Mass., 1962). See also his presidential address to the American Economic History Association, 'Decision making and modern institutional change', *Journal of Economic History*, vol. 33 (1973).

[8] E. T. Penrose, *The Theory of the Growth of the Firm* (Oxford, 1959), p. 3.

[9] The British government now operates a 'thirty-year rule', so that the official documents relating to the interwar period have recently been made available. Some large corporations operate a similar rule, though most have no official policy of allowing access to archives.

The focus of much of what follows differs somewhat from traditional studies in this area, which have been concerned not so much with the growth of the firm as with the implications of monopoly in individual markets. Of course, businessmen have in the past, as now, welcomed the opportunity of strengthening their market position and achieving monopoly powers. It is precisely because of this, however, that the desire for monopoly cannot provide the major explanation of change: the monopoly motive alone has no *diachronic* significance. Businessmen who now enjoy a monopoly no doubt like it, but so presumably would their nineteenth century forebears have liked it. The *historical* problem is to explain why changes in structure occurred when, in the event, they did. Other causal variables must therefore be introduced into the model to explain the changing configuration of firms over time. The technical and institutional background to entrepreneurial decisions is here very important. Government often sets the rules of the game in the market place, and it can be argued that official tolerance of restrictive practices in the past inhibited the natural tendency to merger and concentration.[10] Capital market imperfections, the managerial difficulties of running large organizations, consumer dislike of monopoly, the slow development of the takeover bid, and a widespread preference for small, family controlled enterprises; all these might also act as a significant brake on mergers. When such institutional variables are introduced into the analysis, we will suggest, the steady growth of the large corporation with national market power becomes a more intelligible phenomenon. Whilst the desire for monopoly must be an important element in any model of increasing concentration, then, it has little explanatory value by itself.

The analysis presented here attempts, using economic theory and historical evidence (where these are available in appropriate form), to explain the rise of the corporate economy, which has been such an important part of recent economic history. The aim is to identify systematic factors favouring or retarding the growth of large firms at successive periods in history, with emphasis on the interwar years in which so many of the significant developments occurred. Already, at that time, the possibility of increasing returns in industry leading to a high level of concentration was well understood. Alfred Marshall, for example, drew attention to the point that, with the potentially perpetual

---

[10] Because of factors such as this it may be more correct to see increasing concentration as a change in the kind of monopoly, rather than in the degree of monopoly power exercised. This issue is more fully discussed in Chapter 11.

life of a joint stock company, and increasing returns to scale, there was
almost 'nothing to prevent the concentration in the hands of a single
firm of the whole production of the world'.[11] More recently, economists building on the earlier work of the French statistician Gibrat
have shown that even if there are no systematic tendencies to increasing
returns to scale, concentration may still increase.[12] This possibility –
now known as the 'Gibrat effect' – might be thought of as a process of
natural selection, the result of a situation in which in any period some
firms do well and some firms do badly. This dispersion in performance
over successive periods can induce a steady increase in the dispersion
of firm sizes, and as the successful firms increase their share at the
expense of the unsuccessful ones concentration will increase. Thus
even without any systematic tendency for large firms to experience
more rapid growth than other firms, output will become increasingly
concentrated in the hands of the successful firms.[13] However, the most
recent estimate of the possible impact of the Gibrat effect suggests that
it cannot by itself account for concentration increases of the magnitude
which are in fact observed historically.[14] The bulk of the historical
change in concentration has been due to more systematic factors, and
it is with these – and particularly with the growth of mass production
and the impact of mergers – that we will principally be concerned.

[11] A. Marshall, *Industry and Trade* (1st ed. 1919), p. 316, quoted in S. J. Prais,
'A new look at the growth of industrial concentration', *Oxford Economic
Papers*, vol. 26 (1974).
[12] R. Gibrat, *Les Inégalités Economiques* (Paris, 1931). Prais, 'A new look at the
growth of industrial concentration'.
[13] The same might be true of, say, the distribution of wealth, and in this case
there would be a progressive concentration of wealth in fewer hands, even if
there were no inherent advantages to large rather than small wealth holdings.
However, in the case of wealth, increases in concentration do not in fact occur
because regressive tendencies (the term is used in the Galtonian sense, see
Prais, 'A new look at the growth of industrial concentration'), such as partible
inheritance and estate duties, operate to neutralize it. Of course regressive
tendencies may in some periods neutralize the Gibrat effect's power to increase concentration among firms also (see Chapter 9).
[14] L. Hannah and J. A. Kay, *Concentration in Modern Industry: Theory, Measurement and the UK Experience* (forthcoming 1976).

2

The industrial inheritance: entrepreneurs and the growth of firms to 1914

We have to reckon with the probability . . . one might without
exaggeration say the certainty . . . that we are in the early
stages of the evolution of the form which industry
will take in the future.

H. W. MACROSTY, *The Trust Movement in
British Industry* (1907), pp. 330–1.

కఁ

The growth of large-scale corporations of the modern type is historically related to the process of industrialization, but the link between
the two is neither simple nor direct, for large companies predate
industrialization, and small companies remain common even in mature
industrial societies. Before the onset of the industrial revolution in
England, there were large joint stock overseas trading companies, and
in 'pre-industrial' manufacturing industries also there were large enterprises such as the naval arsenal at Chatham and Whitbread's brewery
in London. Although large-scale firms were not, then, unknown before
the onset of industrialization proper, it was the introduction of new
machine technologies and the application of steam power to manufacturing processes which, from the later eighteenth century onwards,
radically transformed the nature of capitalist enterprise, and created
an economy in which factory concerns employing hundreds (and in
some cases thousands) of workers became the representative form of
business unit.

This transformation in the structure of enterprise was first evident
and advanced most rapidly in the cotton textile industry, which within
several generations changed from a pre-industrial, home based, craft
industry into a highly mechanized, capital intensive factory industry.[1]

[1] See generally, S. D. Chapman, *The Cotton Industry in the Industrial Revolution* (1972).

In the first half of the eighteenth century cotton manufacture was typically organized and financed by merchant-manufacturers who 'put out' the cotton 'wool' to be spun into yarn and woven into cloth by workers in their own homes. The unit of production was small and family based. Typically the work on the hand-operated spinning wheels and looms would be carried out by husband, wife and children working in a domestic setting. Some of the merchants who organized the links between domestic workers, and dealt in the raw materials and final product markets, employed many thousands of individuals and many hundreds of looms; but their direct problems of supervision and management had little in common with those of the later factory masters, and fixed investment per operative was rarely more than a couple of pounds. However, after a series of mechanical innovations in the processing of cotton and with the subsequent application of steam power to these innovations, there was an enhanced need for close control over labour and for centralization of operations to gain full access to economies of scale in the operation of machinery and power equipment. The advantages of factory based production thus increasingly made themselves felt, and the number of mills grew rapidly. By 1822 the representative Manchester cotton mill employed between 100 and 200 operatives, each backed by an average capital investment of between £50 and £90. By the 1830s, new mills containing 40,000 spindles and costing over £80,000 – an increase by a factor of ten on the scale of production familiar a decade earlier – were the order of the day.[2]

It might reasonably be expected that this development would lead to the concentration of production in fewer and larger plants in the hands of a small number of firms. Yet this was not the case, for although the size of firms was increasing rapidly, the growth of the market matched their rate of expansion and in some years even outstripped it. The quantity of raw material processed by the cotton industry, for example, rose by more than 200 times between 1750 and 1850[3] so that, despite a substantial rise in the average size of factories, the market was sufficiently large to accommodate many competing firms at the end of this period, as at the beginning. Many new sources of scale economies were opened up by this expansion of the market, but using Marshall's

[2] Chapman, *The Cotton Industry*, p. 26. H. Perkin, *The Origins of Modern English Society 1780–1880* (1969,) p. 109. S. Pollard, *The Genesis of Modern Management* (Harmondsworth, 1968), pp. 44–7.

[3] B. R. Mitchell and P. Deane, *Abstract of British Historical Statistics* (Cambridge, 1962), pp. 177–9.

celebrated distinction between internal and external economies,[4] we can see that scale economies did not necessarily lead to larger firms. Many of the scale economies were external to the firm. The specialized cotton markets and services which grew up in Lancashire, for example, achieved economies by servicing many firms locally concentrated. Where the scale economies were internal to the firm, they appear to have been exhausted at a scale well below the full extent of the market, so that no firm had a dominant share of output. This pattern of development, favouring the multiplication of small firms rather than the dominance of large ones, continued for most of the nineteenth century. Between 1792 and 1850, for example, the number of factories increased from about 900 to over 1400, with their average throughput increasing by thirteen times over the same period.[5] The modern dilemma – that of enjoying the economies of large-scale production yet preserving the benefits of competition between firms – did not present problems for the nineteenth century cotton industry.

This pattern of development in the textile industry during the industrial revolution was the harbinger of future developments rather than the typical form adopted by contemporary industry. There was a tendency to increasing scale over a wide range of industries,[6] but many others, like the Birmingham metal trades, typically retained a structure of small workshops and enterprises in which the master worked closely with only a few men, a structure both less concentrated and with less dramatic social consequences than the factory system. It was not until the later era of large machine tools, engineering standardization and the assembly line that they were to gain access to internal economies of scale comparable to (and indeed greater than) those in the cotton industry. Other industries, such as pottery, brewing, chemicals and iron smelting (which even before the industrial revolution had not been organized on a domestic basis but rather in plants separate from the workers' homes), did show increasing returns to scale, as a result of technical changes and further divisions of labour during the industrial revolution. Again, however, rapidly expanding markets were

---

[4] A. Marshall, *Principles of Economics* (8th ed. 1920), pp. 221, 237–9, 378–80.

[5] Chapman, *The Cotton Industry*, p. 70. Because of multiple ownership the increase in the number of firms was probably below that in the number of factories, but not significantly so.

[6] For thorough surveys of the contemporary evidence, see: S. Pollard, *The Genesis of Modern Management* (Harmondsworth, 1968), p. 78 ff.; H. Perkin, *The Origins of Modern English Society 1780–1880* (1969), pp. 107 ff.; J. H. Clapham, *An Economic History of Modern Britain*, vol. 1 (Cambridge, 1926), pp. 143 ff.

sufficient to neutralize the effects of such new scale economies, so that competition between many firms was also preserved in these industries. By 1871, then, when over half of the working population were employed in factories (and considerably more than one half of output was produced in factories), it could be argued not only that the degree of competition was no less than it had been previously, but even (given larger markets and greater competition between larger numbers of firms) that it had become more intense.

This is not to say that there were *no* imperfections of competition in the British economy during industrialization. Indeed it may legitimately be doubted whether the idealized state of 'perfect' competition (as conventionally defined in microeconomic theory) existed at this (or indeed at any other) time. Certainly contemporary classical economists nurtured the deep rooted belief that there were widespread restraints on competition in the nineteenth century economy, with harmful effects on welfare. There is evidence of semi-monopolistic price agreements between manufacturers, particularly in the earlier years of the century,[7] and earlier still Adam Smith had provided a salutary warning about the prevalence of informal or secret arrangements:

> We rarely hear [he wrote] ... of the combinations of masters, though frequently of those of workmen. But whoever imagines, upon this account, that masters rarely combine is as ignorant of the world as of the subject. Masters are [he concluded] always and everywhere in a sort of tacit, but constant and uniform combination.[8]

In the main, however, whilst employers did often discuss wages, prices and market conditions, their market power was severely limited by the competitive forces inherent in the nineteenth century industrial structure of many small firms operating in large markets. The efforts of trade associations to control prices were rarely attended with success, and one might almost conclude that the existence of such associations was a symptom of severe competition rather than a serious restraint on it. As the nineteenth century progressed, therefore, contemporary economists principally directed their criticisms not to the only weakly monopolistic arrangements in manufacturing industry, but rather to those

---

[7] E.g. S. R. H. Jones, 'Price associations and competition in the British pin industry 1814–1840', *Economic History Review*, vol. 26 (1973).

[8] Adam Smith, *The Wealth of Nations* (1776) (E. Cannan, ed. 6th ed. 1950), vol. 1, p. 75.

like the chartered overseas trading companies which derived perman-
ence and security from statutory sanction. The widespread dislike of
such monopolies, which were less subject to competitive restraints than
manufacturing firms, stimulated the withdrawal of statutory monopolistic privileges from these companies on a tide of legislation in-
spired by enthusiasm for free trade and *laissez-faire*.[9]

Yet at the same time as such well-established monopolies and restraints of trade were being broken down by legislative and economic
forces, there were independent industrial developments which presented
new monopolistic challenges and created even larger concentrations of
economic power. The railways are the foremost example of this tendency. The railway companies operated on a scale far greater than that
seen in the cotton industry or, indeed, in any nineteenth century
manufacturing industry. Interestingly, it is possible to discern many of
the characteristics of modern corporations in the large railway enterprises of the period.[10] In addition to displaying conventional economies
of scale in operation, they were involved in amalgamations and takeovers to strengthen coordination and spread managerial overheads.
The majority of railway companies were also quoted on the London
and provincial stock exchanges and this promoted the divorce of ownership and control, with shareholders seeing their holding as a *rentier*
investment, leaving the conduct of the business to professional managers
who frequently held only a minority shareholding and pursued different goals. In other respects, however, the railways presented unique
problems different in kind from those encountered in large manufacturing corporations. Their needs in management planning and marketing were rather different and, like similar enterprises in the gas, water
and electricity industries (but unlike most manufacturing corporations),
they were 'natural' monopolies in the sense that uncontrolled competition between many lines running between the same two points would
clearly be grossly wasteful of resources. Although a *modus vivendi* was
devised which partially met these difficulties, the new problems of
regulation which the railways posed fitted uncomfortably in the
Victorian compromise of private enterprise, *laissez-faire* and regulation

[9] P. Mathias, *The First Industrial Nation* (1969), pp. 293–5.
[10] G. Channon, 'A nineteenth century investment decision: the Midland
Railway's London extension', *Economic History Review*, vol. 25 (1972).
M. Zinkin, 'Galbraith and consumer sovereignty', *Journal of Industrial
Economics*, vol. 16 (1967), pp. 3–4. For a more sceptical view of the similarities,
see G. Hawke, *Railways and Economic Growth in England and Wales 1840–
1870* (Oxford, 1970), pp. 384–92.

by competition rather than by the state.[11] Yet, while the railways posed a problem by themselves being potential monopolists, their main impact on manufacturing industry was in the contrary direction of weakening monopoly power. By cheapening transport they promoted the unification of local markets into a national one and, indirectly, by promoting urbanization, they encouraged the standardization of tastes and a trend towards mass production. Thus, whereas in the early stages of the industrial revolution a few manufacturers in a small town might have a local monopoly in their product, by the mid Victorian period they were increasingly having to reckon with goods 'imported' from neighbouring regions on the expanding railway network.

In the 1870s and 1880s, then, after a century of rapid growth both for individual firms and for the economy as a whole, competition in the manufacturing sector was strong, indeed stronger, some would argue, than it had been in any subsequent decade (but see pp. 186–193). Most industries had a multiplicity of what, by modern standards, would be considered small firms, and most consumers faced an expanded choice between the goods of a larger number of sellers from a wider area. Product differentiation enabled a few manufacturers of specialized products to obtain a favoured market position[12] and other producers enjoyed a local market protected by local tastes or long distances from other potential competitors. In general, however, such factors mitigating the severity of competition were rare. In almost all industries demand was growing rapidly and, in many, more rapidly than the size of firms. Thus, despite their explosive nineteenth century growth, it is unlikely that the largest 100 firms in 1880 accounted for even as much as 10 per cent of the market, compared with something nearer 50 per cent which they control today (see Appendix 2).

Yet these very conditions of competition and rapidly expanding markets contained within them the impetus to the division of labour which in the long run was to result in the greater concentration of output in the hands of large firms. Towards the end of the nineteenth century it became generally recognized that these tendencies to industrial concentration were becoming increasingly marked. This development was perhaps in part a natural result of the continued growth of firms, coupled with the slower expansion of the market as the pace of

[11] H. Parris, *Government and the Railways in Nineteenth Century Britain* (1965).
[12] E.g. P. L. Payne, *Rubber and Railways in the Nineteenth Century* (Liverpool, 1961), pp. 95–113.

economic growth in Britain slowed down.[13] It was also the result of a
process which had hitherto had little impact on competition but which
had in fact been set in train at a very early stage of the industrial
revolution. The essence of this process was sketched by Adam Smith
in the early chapters of the *Wealth of Nations*, and encapsulated in the
famous phrase: 'the division of labour is limited by the extent of
the market'.[14] Sustained growth was possible in Smith's model of
economic evolution because not only did expanding markets increase
the opportunities for divisions of labour, but each division of labour
in its turn raised productivity and thus made possible a further expansion of markets, bringing the process into a self-sustaining virtuous
spiral of progress. Competition, in Smith's model, provides a spur to
firms to achieve economies of scale and specialization. An increase in
demand, he argues, 'encourages production, and thereby increases the
competition of the producers, who, in order to undersell one another,
have recourse to new divisions of labour and new improvements of
art, which might never otherwise have been thought of'.[15] The contrast
between this view and the concept of competition in the equilibrium
models of later economic theory is noteworthy. In the modern formal
models possible divisions of labour are assumed to be exhausted and
the appearance of further economies of scale would be inconsistent with
the continuance of competition. In Smith's dynamic model of economic
evolution, also, the question arises of how competition will be maintained in the presence of increasing returns to scale, though Smith
himself does not appear either to have been troubled by this problem
or to have resolved it. There are a number of possible explanations of
the failure of such long run tendencies to materialize.[16] We have already
noted one of these – the growth of the market itself. An increase in
market demand must be met either by the further expansion of existing
firms or by the entry of new ones. The precise balance between the
contribution of the methods will depend on the degree to which the

13 P. Deane ('New estimates of gross national product for the United Kingdom
   1830–1914', *The Review of Income and Wealth*, vol. 14 (1968), p. 96) suggests
   that GNP grew at 2.4 per cent per annum between the 1840s and 1870s, but
   at only 1.9 per cent per annum between the 1870s and 1914.
14 Smith, *Wealth of Nations* (Cannan, ed.), vol. 1, p. 21. See also: A. Young,
   'Increasing returns and economic progress', *Economic Journal*, vol. 38 (1928);
   G. B. Richardson, 'Adam Smith on competition and increasing returns',
   paper read at Adam Smith conference, Kirkcaldy, 1973.
15 Smith, *Wealth of Nations* (Cannan, ed.), vol. 2, pp. 271–2.
16 See Richardson, 'Adam Smith on competition and increasing returns', for a
   fuller consideration of the possible explanations.

newly enlarged market induces new internal economies of scale. There can be no general presumption, at least from Smith's imprecise and intuitive model, that the magnitude of such internal economies will exceed the extent of market growth. Indeed, as we have seen in the cotton industry, the evidence suggests that technical progress and the division of labour proceeded rapidly but not as rapidly as the growth of the market. Hence the expansion of production was achieved by a combination of the growth of existing firms and the entry of new ones.

The maintenance and strengthening of competition by the new entrants in the cotton industry was greatly facilitated by the emergence of a specialized engineering industry, a further important 'division of labour and improvement of art' induced by market expansion. Specialist machinery builders would supply factory equipment incorporating the latest designs of, say, power looms and the steam boilers and engines to drive them, and this encouraged the entry of new firms on equal terms with existing ones. Since the machine building industry was itself competitive such equipment was freely available on the open market. Yet there could be no presumption that this would continue indefinitely, and experience soon showed that control over machine technology could evolve differently. As competition among machine builders and users stimulated further inventions and division of labour, the inventors and machine builders perceived an alternative strategy which promised them higher financial returns. The essence of this was that a firm which could reserve the use of a new or significantly improved type of machine to itself could expect to use this monopoly to grow more rapidly than its competitors and gain a market share larger than would be necessary purely to achieve any economies of scale inherent in the equipment itself.

An example of such a change drawn from the tobacco industry will perhaps make the opportunities opened up by this alternative strategy clearer. Tobacco was traditionally sold in loose form, but by the mid Victorian period rolled cigarettes were becoming increasingly esteemed and, in the manner of Smith's model, an expansion of the market for such cigarettes stimulated a search for machines which could roll them efficiently. An important element in the success of the firm of W. D. & H. O. Wills (whose branded cigarettes were market leaders) was that they acquired exclusive control of one such invention, the Bonsack cigarette machine, a US invention which could produce from 300 to 500 cigarettes a minute. For this machine, neither capital costs nor operating economies of scale were large in relation to the rapidly expanding market, but, by improving on the basic machine and by buying out the patents of its rivals, Wills were able to gain an important competitive advantage over other producers. Demand for their mass-produced penny cigarettes, such as Woodbine, increased dramatically, and by 1901 Wills's share of cigarette sales was as high as 55 per cent of the UK total. That this offered substantial scope for monopoly profit is confirmed by Wills's rate of return of 62 per cent on capital employed in that year, four times their rate of profit of two decades earlier.[17]

The advantages of such a near monopolistic position, conferred by an assured lead in a specialized machine technology, were perceived in other industries,[18] but there were also many markets in which dominance was achieved not by restricting the use of technology but simply because the technology developed was efficient only at such a large scale that a small number of plants could satisfy the whole of the market demand which could reasonably be anticipated. For this reason also, therefore, firms began to expand at a pace which outstripped the rate of growth of markets. This development was accelerated in the closing decades of the nineteenth century by significant changes in the nature of market demand for some consumer goods. As urbanization progressed, and urban incomes rose, consumer tastes for a range of common products became more standardized and the possibilities in large cities of scale economies in marketing were increased.[19] There were, for example, economies both in advertising branded goods, and in establishing sales forces to promote them through the multiplicity of highly competitive retail outlets. These marketing economies also had repercussions on production economies of scale. Where demand for a product could be standardized, and perhaps also widened, the short and specialized production runs which had been necessary to cater for more sharply differentiated local tastes could be replaced by longer runs and often by more economical and more capital intensive processes. These

[17] For a fuller account see B. W. E. Alford, *W. D. and H. O. Wills and the Development of the United Kingdom Tobacco Industry 1786–1965* (1973), pp. 139–57, 223, 225–33, 302.

[18] Examples are numerous in the reports of the Monopolies Commission and in business histories, though curiously the phenomenon has been generally ignored by economic historians. It would repay more systematic investigation, for, like the patent system on which it is sometimes based, its effect may be either beneficial (in encouraging invention) or harmful (in suppressing innovation).

[19] A. F. Weber, *The Growth of Cities in the Nineteenth Century* (New York, 1899, reprinted 1965), pp. 40–64.

tendencies were not by any means a new phenomenon, but they do appear to have intensified from the 1880s onwards. Significantly, it was in industries which experienced such a shift in the pattern of demand – cigarettes, wallpaper, flour, soap, sewing cotton and linoleum – that tendencies to higher concentration were most conspicuous.[20] Even in the brewing industry, where the emergence of a mass urban market had stimulated the growth of firms at a much earlier period, the tendencies to higher concentration were given a further impetus by new marketing developments, as local magistrates after 1869 attempted to limit the number of public houses by reintroducing a restrictive licensing system. This stimulated a scramble to acquire small breweries and groups of public houses as the big brewers sought to consolidate their hold on the mass urban markets.[21]

In a significant number of industries, then, firms in the closing decades of the nineteenth century were facing important shifts both in the technical basis of production and in the nature of market demand. In industries where this resulted in the emergence of economies of scale in production and marketing, there was (given that the majority of them contained a large number of vigorously competing firms) an inevitable tendency to overextension of capacity and price cutting.[22] If bankruptcy for all were to be avoided there were a number of possible responses to this situation. In some industries leading firms extended their plant to the new optimum efficient size, and by reducing prices to below the marginal cost of the earlier technology they eliminated many of their rivals. In others the majority of existing firms renewed their attempts at collective price and output control. Such combinations were, however, inherently unstable, for, although there was a general interest in the controls, individual members had a powerful incentive to maximize their revenue by expanding production secretly and selling below the agreed price. One possible response to this realization was a more permanent and watertight merger of interests. In contrast to a cartel, a merger of this kind had the advantage that the otherwise divergent interests of the contracting parties were closely and formally cemented together into a common cause, and the individual firm's incentive to renege on the agreement was eliminated.

[20] H. W. Macrosty, *The Trust Movement in British Industry* (1907).
[21] J. Vaizey, 'The brewing industry', in P. L. Cook and R. Cohen (eds), *The Effects of Mergers* (1958), pp. 400–11.
[22] W. Ashworth, *An Economic History of England 1870–1939* (1960), pp. 91–102. J. S. Jeans, *Trusts, Pools and Corners* (1894), pp. 3–4. H. W. Macrosty, 'The grain milling industry', *Economic Journal*, vol. 12 (1903).

Of course, the possibility of a full merger of interests had been appreciated in earlier decades of the nineteenth century. Indeed, merger, viewed simply as the combination of businesses, is at least as old as marriage, and alliances of family business interests cemented by marriage and kinship remained a major source of additional finance and business expansion throughout the nineteenth century. The division of firms between heirs on the death of the owner was also practised, underlining the essentially personal and familial nature of business alliances (a contingent result of such division was to check the tendency of marriages to increase the size of businesses). Despite the constraints of this familial framework, it was possible for a small number of outstanding businesses to grow to a considerable size. As early as 1795 the Peels (then the largest business in the cotton industry) owned twenty-three mills in Blackburn, Bury, Bolton, Burton-on-Trent and Tamworth; and many other entrepreneurs, in this and other industries, extended their control over many factories as they reinvested profits and gained competitive advantages over rivals which enabled them to buy them out.[23] Increasingly, capital needs for such expansion were also met by accepting money on loan from the general public and also by admitting partners – whether 'sleeping' or otherwise – to formal participation in the enterprise. In some industries where capital requirements were large, the partnership was the normal business form and British entrepreneurs had not in general been averse to diluting family ownership by merger in this way where it was economically advantageous to do so.[24]

However, it seems clear that before such *ad hoc* partnerships could develop further into a more modern form of corporate enterprise, institutional developments both in company law and in stock exchange practice were necessary. Formally, the required legal changes came in England and Wales between 1844 and 1856 when first joint stock companies and then limited liability companies received the general sanction of parliament.[25] These had already been available under

[23] Chapman, *The Cotton Industry*, pp. 29, 32. Alford, *Wills and the Development of the UK Tobacco Industry*, p. 73.

[24] Chapman, *The Cotton Industry*, pp. 38–9. D. Landes, *The Unbound Prometheus* (Cambridge, 1969), pp. 72–3.

[25] The paragraphs which follow are based on: H. A. Shannon, 'The coming of general limited liability', *Economic History*, vol. 2 (1931); H. A. Shannon, 'The first 5000 limited companies and their duration', *Economic History*, vol. 2 (1932); H. A. Shannon, 'The limited companies of 1866–1883', *Economic History Review*, vol. 4 (1933); G. Todd, 'Some aspects of joint stock companies 1844–1900', *Economic History Review*, vol. 4 (1932); J. B.

separate legislation for public utilities or by private act of parliament, but by these new acts the way was opened for a relatively cheap company form for manufacturing industry also. The facility of joint stock was more convenient than either a full partnership or the deposit of money with a firm, since the shares were readily transferable and the rights of control which they carried strengthened the security of the investor. Finally, with the granting of limited liability, the joint stock investor was also relieved of the responsibility for the whole of the debts of the firms in which he held shares. In retrospect, the simple device of making him liable only up to the value of shares for which he undertook to subscribe seems to be both realistic and just. It was a natural extension of the similar rights which had been accorded to investors in, for example, railways, and reduced the risks of *rentier* investors in businesses which were in reality run not by them but by the directors.

Yet this favourable view of the new company form was by no means shared by all contemporaries. Indeed the balance of public opinion (as expressed to the numerous parliamentary committees which discussed the merits and demerits of the proposals) reacted unfavourably to the idea of allowing shareholders in a business to repudiate responsibility for its debts and foist part of their risks on to suppliers and customers. Even those who accepted the arguments in favour of limited liability – principally its favourable effects on investment – saw little future for the new companies, since, as they were forced to advertise their limited liability status by the tag 'Limited', suppliers and customers would, it was thought, be reluctant to deal with them. There were also serious doubts about whether the management of such companies would be efficient and reliable, and the view of Adam Smith was still commonly held:

> The directors of such companies [he wrote] . . . being the managers rather of other people's money than of their own, it cannot be well expected, that they should watch over it with the same anxious vigilance with which the partners in a private copartnery fre-

Jefferys, 'The denomination and character of shares 1855–1885', *Economic History Review*, vol. 16 (1946); J. B. Jefferys, *Trends in Business Organisation in Great Britain since 1856*, unpublished PhD thesis (London, 1938); (Balfour) Committee on Industry and Trade, *Factors in Industrial and Commercial Efficiency* (1927), pp. 125–7; Sir John Clapham, *An Economic History of Modern Britain*, vol. 1 (1938), pp. 201–91. See also R. H. Campbell, 'The law and the joint stock company in Scotland', in P. L. Payne (ed.), *Studies in Scottish Business History* (1967).

quently watch over their own. ... Negligence and profusion, therefore, must always prevail, more or less, in the management of the affairs of such a company.[26]

This view was rooted in the direct experience of businessmen as well as in inductive reasoning. The elder Sir Robert Peel, for example, thinking of the problems of large-scale management, felt that, 'It is impossible for a mill at any distance to be managed unless it is under the direction of a partner or superintendent who has an interest in the success of the business.'[27] Moreover, the early years of limited liability abundantly confirmed that these contemporary fears and prognostications were not entirely without foundation. Over 30 per cent of the public companies formed between the achievement of general limited liability in 1856, and 1883, ended in insolvency, many of them in the first five years of their existence. There was a strong suspicion that this rate of failure resulted from the activities of crooked promoters out for a quick speculative profit, a view lightheartedly reflected in W. S. Gilbert's *Utopia Limited*:

> Some seven men form an Association
> (If possible all Peers and Baronets)[28]
> They start off with a public declaration
> To what extent they mean to pay their debts
> That's called their Capital . . .[29]

The early failures deepened distrust among the public and doubts were strengthened also by the multiplication of unexpected calls on partly paid shares (a form of share which was then common, though it eroded the limitation of liability). As a result, by the 1880s perhaps only 5 to 10 per cent of those larger-scale businesses which might have been expected to benefit from outside investment had in fact adopted limited liability company status.

Already by that time, however, there were improvements in the practices of limited liability companies which were to give them wider popularity. Smaller uncalled liabilities and lower share denominations

26 Smith, *Wealth of Nations* (Cannan, ed.), vol. 2, pp. 264–5.
27 *Parliamentary Papers*, 1816, III, *S.C. on Children Employed in the Manufactories of Great Britain*, p. 136, quoted in Perkin, *Origins of Modern English Society*, p. 114.
28 A reference to the practice of recruiting 'guinea pig' directors to lend lustre to a company prospectus.
29 W. S. Gilbert, *Utopia Limited*, in *Original Plays*, Third series (1895 ed.), p. 434.

became more common and this, together with the development of under-writing, led to a wider market in industrial issues. Solicitors, stockbrokers, company agents, bankers and accountants acted as intermediaries with the investing public, with whom they were closely in touch; and large firms in industries such as iron and steel and shipbuilding began to see benefits in raising their additional capital needs through these channels. Public demand for issues of shares in manufacturing industries was also boosted by some popular and oversubscribed issues in the consumer goods field. In 1886, for example, Guinness was floated as a public company and in 1890 there followed the flotation of the sewing cotton combine J. & P. Coats. There was then a rapid and continuous increase in the paid up capital of companies and many of these companies offered their shares to the public. Between 1885 and 1907 the number of firms in domestic manufacturing and distribution with quotations on the London stock exchange grew from only sixty to almost 600,[30] and the provincial stock exchanges 'were almost of greater importance in relation to home securities than London'.[31]

This development of the stock market not only facilitated the growth of firms facing the pressures for economies of scale in marketing and production, but also created new economies of scale in financing, and thus added to these pressures. Many of the companies which might have wished to seek capital on the provincial or metropolitan stock markets were too small to raise money economically themselves.[32] They could, however, join together with other firms and make a joint issue. The most natural way of doing this was, of course, to make a public flotation of the businesses of the would-be borrowers as a merged company, and access to economies of scale in the capital market of this kind appears to have been the aim behind many of the mergers of these years. It has also been suggested[33] that these developments in company financing created a new financial impetus to the merger movement by encouraging speculative activity in the stock market. Investors were relatively inexperienced in assessing the shares of manufacturing firms and their profit expectations were rather insecurely based. It was a

[30] P. E. Hart and S. J. Prais, 'The analysis of business concentration: a statistical approach', *Journal of the Royal Statistical Society*, series A, vol. 119 (1956), p. 154.
[31] Quoted from G. H. Phillips, *Phillips' Investors Annual* (1887), in Jefferys, *Trends in Business Organisation*, p. 340.
[32] F. Lavington, *The English Capital Market* (1921), p. 219.
[33] L. Hannah, 'Mergers in British manufacturing industry 1880–1918', *Oxford Economic Papers*, vol. 26 (1974).

B

simple matter for an unscrupulous company promoter to inflate these
expectations by issuing an optimistic prospectus, for neither the state
nor the Stock Exchange Committee provided any real safeguards for
the inexperienced investor. The potentially large benefits of economies
of scale and monopoly profits which could be promised in a merger
could be particularly appealing in the new issue market. First, however,
the promoter had to convince the present owners of the companies
(whose profit expectations would be tempered by experience and close
personal acquaintance with the business) that the sale of their businesses
to the public could be lucrative. To do this it was necessary to offer
them a price which significantly exceeded the discounted present value
of their own more securely based profit expectations. The evidence
suggests that a number of promoters could successfully achieve this.
Frank Harris, for example, was 'admirably received'[34] by the owners of
the Bovril Company when it was known that he was acting for Ernest
Terah Hooley, a promoter with a series of successful promotions to his
credit. Harris claims that the owners wanted £1½ million for their
business and that he offered £1¼ million but was beaten by another
agent of Hooley's who offered £2 million! Such offers could, of course,
only be based on Hooley's masterly capacity for exaggerating share
values when at a later stage he sold the company through a public issue
of shares. Potential shareholders seemed most ready to accept such
inflation of values beyond what private businessmen would reasonably
anticipate, when the general share price level had been rising and was
nearing its peak: investment success in the boom bred speculative
activity, at least until the bubble burst. Significantly the majority of
merger issues by promoting syndicates were made precisely in such
boom periods. The positive correlation between the level of share prices
and the intensity of merger activity[35] is consistent with this hypothesis
that financial factors played an important part in stimulating the
amalgamations of this period. This view is also confirmed by the complaints of 'overcapitalization' which almost invariably followed the
more unscrupulous or misguided merger issues.[36] Such complaints
clearly suggest that the profits which were made subsequently to an
issue were insufficient to service the large amounts of capital which had
been subscribed.

[34] F. Harris, *My Life and Loves* (1966 ed.), p. 827.
[35] Hannah, 'Mergers in British manufacturing industry 1880–1918', p. 9, using
  data for 1880–1918, records an $\bar{R}^2$ of 0·62 in correlating share prices and
  merger activity.
[36] Macrosty, *The Trust Movement in British Industry*.

Thus on a number of fronts – technical, commercial and also financial – there were, in the closing decades of the nineteenth century, both strong pressures and new opportunities making for larger-scale enterprise. The result was that the merger waves of these years were far more intense than any which had been experienced earlier in the century.[37] Between 1888 and 1914 an average of at least sixty-seven firms disappeared in mergers in each year, and in the three peak years of high share prices and intense merger activity between 1898 and 1900 as many as 650 firms valued at a total of £42 million were absorbed in 198 separate mergers. This important turn of the century merger wave was particularly marked in the important textile finishing industry, in which a series of combines, with control over an average of 80 per cent of their respective markets, were floated: the Bradford Dyers' Association (1898, accounting for twenty-one firm disappearances), the Calico Printers' Association (1899, forty-five firms), the Bleachers' Association (1900, fifty-two firms) and the British Cotton and Wool Dyers' Association (1900, forty-six firms). There were further large mergers in other branches of the textile industry: the Fine Cotton Spinners' and Doublers' Association (1898, thirty firms) and the Yorkshire Woolcombers' Association (1899, thirty-seven firms). In other industries also there were consolidations in this period, the more important among them being British Oil and Cake Mills (1899, sixteen firms), Wallpaper Manufacturers (1900, thirty firms), Imperial Tobacco (1901, twelve firms) and the Associated Portland Cement Manufacturers Company with its later subsidiary British Portland Cement (1900 and 1912, fifty-nine firms).[38] Almost all of these claimed, and many of them achieved, market shares of between 60 per cent and 90 per cent. Such large-scale multi-firm mergers were spectacular and gained as much attention from contemporaries as they have subsequently from historians; but scarcely less important were the less noticeable mergers that were also becoming commonplace. The larger firms in the brewing industry acquired several hundred smaller breweries in these years, and large

---

[37] The account of the merger waves which follows draws substantially on: Hannah, 'Mergers in British manufacturing industry 1880–1918'; Macrosty, *The Trust Movement in British Industry*; and P. L. Payne, 'The emergence of the large-scale company in Great Britain, 1870–1914', *Economic History Review*, vol. 20 (1967).

[38] Acquisitions of foreign or non-manufacturing companies are excluded from the figures in this paragraph, so that the total number of firms involved in these mergers was in some cases even larger than indicated. For a fuller listing, see M. A. Utton, 'Some features of the early merger movements in British manufacturing industry', *Business History*, vol. 14 (1972), p. 53.

firms like Vickers in the engineering industry and Levers in the soap industry were also growing rapidly both by internal growth and by merger.

This movement towards industrial concentration was historically unprecedented and it created manufacturing enterprises with capitals distinctly larger than the early nineteenth century cotton lords could have aspired to. The movement in Britain was not, however, on as large a scale as the contemporaneous movement which was occurring for similar reasons in the United States, and it seems certain that in 1914 the level of industrial concentration in the United Kingdom was still below that in the United States.[39] The peak year for mergers in the US was, as in the UK, 1899, but in both numbers and values the US merger wave far outpaced that in this country: in 1899 alone there were 979 firm disappearances by merger valued at \$2064 million (over £400 million),[40] compared with 255 firm disappearances with a value of only £22 million in the UK. There is also evidence that the UK turn of the century merger wave affected fewer industries and created fewer large corporations with smaller market shares than their US counterparts.[41] The contrasting experience of the two countries is surprising, since they faced very similar technologies and were, arguably, at a comparable stage of industrial development. Hence a number of hypotheses have been advanced in explanation of the contrast. US management may have more successfully coped with the problems of greatly enlarged company size.[42] Imperfections in the US capital market coupled with exceptional requirements for financial mobilization may have strengthened the hands of US financiers, like the Rockefellers and Morgans, in creating mergers.[43] Finally, Professor Payne has advanced a multicausal explanation in terms of Britain's early start and technological base, the size and structure of the market for British goods, and the pride in self-sufficiency and anti-professionalism of the British industrialist.[44] None of these hypotheses are, by themselves, entirely con-

[39] L. Davis, 'The capital markets and industrial concentration: the US and the UK, a comparative study', *Economic History Review*, vol. 19 (1966).

[40] R. L. Nelson, *Merger Movements in American Industry 1895–1956* (Princeton, 1959), pp. 139–53. There are some difficulties in making such direct international comparisons, but they are unlikely by themselves to account for such a large discrepancy.

[41] Hannah, 'Mergers in British manufacturing industry 1880–1918', pp. 10–12.

[42] J. S. Jeans (ed.), *American Industrial Conditions and Competition* (1902), pp. 74–85, and see pp. 79–100 below.

[43] Davis, 'The capital markets and industrial concentration'.

[44] Payne, 'Emergence of the large-scale company', pp. 523–7, 533, 536–9.

vincing, though together they pose a formidable programme for further research, if no settled conclusions.

Whatever the reasons for the contrasting experience of the two countries, the results of these industrial developments in the UK are clear. By the decade before the First World War the structure of British manufacturing industry was already very different from that of three or four decades earlier. The foundations of the modern corporate economy were already discernible in the large firms that had been created, with many of the important companies of today (including Imperial Tobacco, Watney, Dunlop, GKN, and Vickers) already being established among the leading firms. If we look more carefully at the largest firms[45] and at their relationship with the rest of industry, however, it is clear that whilst many of the outlines of modern industrial structure appear there, there are also many innovations we now associate with the corporate economy which were then barely apparent. It is true that the quotation of large manufacturing companies on the stock exchange and the widespread merger activity of these years were important innovations. Yet the industrial partnership and the family owned factory remained the typical unit in most branches of manufacturing. The institutional innovations in company law, which strengthened tendencies to large scale, had also given a new lease of life to many smaller businesses. Partnerships and family firms adapted the new institutional form to their own purpose, and by 1914 as many as four-fifths of the registered joint stock companies were private rather than public companies.[46] Even amongst the public companies, moreover, the original family controllers frequently remained in leading managerial positions. It was a common practice, encouraged by the better company promoters, for the founding families to retain a major interest in a newly merged and publicly floated enterprise. This practice had a lot to commend it, for the separation and professionalization of management which we associate with modern corporations still had a long way to develop, and outside managers were not easily come by. Significantly, many of the firms which did not retain the services of the former owners (or which retained the services of owners lacking the managerial capacity to exercise overall control in an enlarged company) encountered serious managerial difficulties (see pp. 82–3). By contrast the three largest companies of the day – which were also among the most

[45] For a list of the fifty largest manufacturing firms of 1905, see Payne, 'Emergence of the large-scale company', pp. 539–40.
[46] (Balfour) Committee, Factors in Industrial and Commercial Efficiency, p. 125.

successful – J. & P. Coats, Imperial Tobacco, and Watney Combe Reid – were not built up around a new corporate management at all, but around old family firms with their senior management and directors recruited principally from among the founding families.[47] While this solved a fundamental problem of the corporate economy – that of maintaining managerial efficiency while divorcing ownership from control – it did so more by avoiding the issue than by devising new techniques of incentive and control.

In other respects also the large firms in the decade before the First World War were very different from those that were to emerge in the interwar period. The numbers of firms surviving is some indication of this – only twenty-two of the fifty largest firms in 1905 were still among the fifty largest in 1930[48] – but the contrast is even clearer in qualitative terms. Many of the large firms of 1905 were in a limited number of traditional industries: the brewing trade, for example, accounted for more than a third of them. Now the brewing industry was a very large one and this meant that the limitation of competition which we would normally associate with such a concentration of big firms was less likely to materialize in this industry. The typical large brewery still met significant competition not only from other large breweries in the metropolitan areas but from small and medium-sized local brewers throughout the country. Moreover, breweries that attempted to charge monopolistic prices would still meet some resistance from untied houses or other public houses on short contract terms. There were, of course, some large companies which more clearly dominated their markets – including Imperial Tobacco, Associated Portland Cement, and the textile finishing combines – but they were the exception rather than the rule among the largest companies in this period.

Furthermore, many of these prewar large firms were not involved in the modern sectors of the economy which we now associate with large corporations. Although many of them were prosperous at the time, they were principally engaged in the staple industries, which were shortly

[47] The *Stock Exchange Year Book* for 1913 shows that thirty-two out of the fifty-one directors of these companies bore the names of their founding family firms. Other rapidly growing firms such as Lever Brothers were controlled by individual entrepreneurs like William Lever rather than by professional corporate managers. See, generally, Payne, 'Emergence of the large-scale company', pp. 530–6.

[48] Compare the list in Payne, 'Emergence of the large-scale company', pp. 539–540, with the list on p. 118 below. However, contrasting measures of size in the two studies make precise comparison hazardous.

to face the strain of war and economic depression, and for many of them this was to mean a decline in their traditional markets. By contrast, the new and rapidly growing industries were poorly represented. There were only three chemical firms (two of them in stagnant or declining sectors of the chemical industry), and none in motor manufacture or electrical engineering, among the top firms of 1905.[49] Of the chemical firms, the Salt Union and the United Alkali Company had been formed in 1888 and 1890 respectively, as defensive moves in traditional sectors of the chemical industry. Only Brunner Mond, using the Solvay process of alkali manufacture, represented a more recent development.[50] No British motor car manufacturer had by then embarked upon techniques of assembly line production, which Henry Ford was at that time successfully introducing in America; and in electrical machinery the British market was dominated by subsidiaries of American and German parents and by imports from abroad.[51] There was, then, little sign of mass production techniques in the new industries, yet it was on these that the capacity to grow of the indigenous science based corporations of the future was to rest.

It has been credibly argued that, during the general industrial stagnation of the years before 1914, Britain experienced a decline in industrial productivity and the lowest recorded rate of economic growth in her history.[52] The coincidence of this retardation of growth with the relative 'backwardness' in merger activity and industrial concentration raises a number of questions. If there is a causal link between the two phenomena, it is, however, difficult in the present state of knowledge to determine in which direction the relationship might lie. It might, for example, be argued that the new industries (on which the 'missing' growth would have been based) were typically highly concentrated, so that the low level of concentration in Britain was simply a function of her retarded economic growth. An alternative view would be that large corporations of the modern type would have been able to induce the adjustment in the structure of output (and particularly the growth of new industries) more efficiently than the multiplicity of

[49] Payne, 'Emergence of the large-scale company', pp. 539–40.
[50] All three firms were eventually to join Imperial Chemical Industries.
[51] G. Maxcy and A. Silberston, *The Motor Industry* (1959), p. 12. I. C. R. Byatt, 'Electrical products', in D. H. Aldcroft (ed.), *The Development of British Industry and Foreign Competition 1875–1914* (1968), pp. 244–73.
[52] See e.g.: E. H. Phelps Brown and Margaret H. Browne, *A Century of Pay* (1968), pp. 174–95; D. N. McCloskey, 'Did Victorian Britain fail?', *Economic History Review*, vol. 33 (1970).

small and medium-sized firms which the British economy had inherited. As contemporaries became more deeply concerned about their overall economic performance, they were more and more inclined to adopt this second view. Increasingly they saw the emulation of the industrial structure of the United States and Germany as the key to revitalizing the British economy. It was in self-examination in this critical vein that the 'rationalization' movement of the interwar years was to take root.

3

The rationalization movement

> The rapid development of the idea of rationalization has
> given rise to amalgamations at a speed and to a
> degree which are altogether novel.
>
> L. F. URWICK, address to the Economics section
> of the British Association for the
> Advancement of Science (1930).

ನಾನಾ

In the United Kingdom the First World War marked a watershed in economic and business development as well as in political and social life. On this much there is general agreement, but historians are divided on the related issue of whether the changes of the war years were uniquely a result of the war itself, or whether the same changes would in any case very soon have arisen, since they stemmed from growing social and economic pressures which were already evident in prewar Britain.[1] Already in the late nineteenth century and the first decade of the twentieth, there were, as we have seen, strong inducements making for larger scale enterprise, and it would not be difficult to trace ideas about industrial organization which later came to be dubbed 'rationalization' to an origin in the amalgamation movements of the prewar years.[2] Yet only during the war and in the years that followed did such ideas begin to attain wide currency and gain the status of the conventional wisdom in leading business circles. The contrast between the pre- and postwar periods in industry was a real one. Churchill's much quoted hope in 1914, that the war would see 'business as usual',[3] was in fact to prove to be profoundly wrong, for between 1914 and 1918 the pattern of production over a wide range of industries was to be transformed as businesses strove to meet wartime requirements. Many materials formerly imported from the Continent were cut off by the

[1] See, generally, A. S. Milward, *The Economic Effects of the World Wars on Britain* (1970).
[2] E.g. H. W. Macrosty, *The Trust Movement in British Industry* (1907), p. 334; A. L. Levine, *Industrial Retardation in Britain 1880–1914* (1967), p. 44–54.
[3] Speech at the Guildhall, Nov. 1914.

war, so that substantial growth of home industries producing products such as magnetos (for vehicles and aircraft) and dyestuffs (for the textile industry) was necessary. However, private initiatives (both in such new industries and in the established armaments industries) soon proved inadequate, and the government, and especially the Ministry of Munitions, began to play an increasing role in planning, financing and directing the activities of manufacturing firms. The state, on a war footing, with a large and predictable demand for a wide range of weapons and general supplies, was in a unique position to influence business firms. In general, it used this position to induce them to adopt the most economical methods, particularly by discouraging product differentiation and by encouraging investment in long runs and mass production to meet government demand. Indirectly this resulted in more merger activity, as firms made acquisitions to expand their capacity; and mergers were further accelerated towards the end of the war, as firms like Vickers and Nobel (which faced a decline of demand for their military products in peacetime) strengthened their position by diversification and acquisition. The internal practices of firms were also profoundly changed by the war: interchangeable standardized parts were increasingly used in the engineering trades; capital and un-skilled labour took the place of skilled craftsmen; and there was in-creasing reliance on electricity as a source of power – during the war years electricity consumption doubled.[4]

Many of these wartime industrial developments – mergers, larger scale enterprise, new industries, standardization and mass production – were, of course, associated with the stirrings of what became the modern corporate economy. They were not forgotten after the war, for once the profitability of the new methods had been realized their impact was cumulative. The industrialist Sir Vincent Caillard typified the indus-trial mood when he looked back from the 1920s to the war years as a time of:

cooperation on a marvellous scale when ... manufacturers for the good of their country, threw away their old prejudices and put them-selves unreservedly at the disposal of one another. Patents, secret

4 E. M. H. Lloyd, *Experiments in State Control* (Oxford, 1924). J. M. Rees, *Trusts in British Industry 1914–1921. A Study of Recent Developments in Business Organisation* (1922). S. Pollard, *The Development of the British Economy 1914–1950* (1962), pp. 42–62, 76–87.

processes, special methods, goodwill, were flung into the melting pot of the common weal.[5]

This was the authentic, exhilarated voice of a movement in business opinion which was growing in strength and which questioned the virtues of competition and championed the advantages of cooperation, merger and large-scale enterprise. Whilst many other wartime changes were abandoned,[6] the changes in business opinion which had accompanied them were, in general, more permanent. Indeed already, during the war, various committees sponsored by the government to inquire into wartime changes and postwar prospects betrayed a taste for continuing many of the wartime innovations in peacetime.[7] The war, so the contemporary optimism ran, had caused a 'New Industrial Revolution which is likely to have far greater results for good than the introduction of machinery 130 years ago'.[8]

Yet the situation faced by British industrial firms after 1919 was profoundly different from that which had been envisaged in the closing stages of the war. Instead of the anticipated postwar prosperity there was, after a brief inflationary boom, a fierce slump, and in 1921 the unemployment rate rose above 10 per cent, never to fall below that figure again before the Second World War.[9] Economic depression was, then, a central fact of the interwar experience: to the labourer it meant the dole, to the employer it meant overcapacity; for both it provoked a further re-evaluation of their political, social and economic beliefs and of the economic institutions they sustained. The effect of the depression in increasing political consciousness and stimulating the growth of the Labour Party, and its impact on the collapse of the classical paradigms of economic theory in the Keynesian revolution, have often been analysed by historians. Less well covered, but no less important, is its impact at the level of popular business philosophy, as more businessmen began to question the desirability of the configuration of firms and markets which they had inherited from the prewar era. The rationaliza-

[5] Quoted, from Caillard's pamphlet *Industry and Production*, by A. Wright, 'The new phase in industry', *Financial Review of Reviews*, vol. 22 (1929), pp. 42–9.

[6] R. H. Tawney, 'The postwar abandonment of economic controls 1919–1921', *Economic History Review*, vol. 13 (1943).

[7] See, e.g., Board of Trade Committee on the Electrical Trades after the War, *Report* (Cmd. 9072, 1918).

[8] Industrial Reconstruction Council, *Reconstruction Handbook* (1918), p. 2.

[9] B. R. Mitchell and P. Deane, *Abstract of British Historical Statistics* (Cambridge, 1962), pp. 65, 67.

tion movement – which gained the attention of bankers, politicians and trade unionists, as well as of prominent industrialists between the wars – was an important aspect of the build up of dissatisfaction with the market mechanism and of the movement towards greater reliance on large firms for economic organization. Though the Keynesian revolution ultimately demonstrated that the malfunctioning which they witnessed also had a macro-solution, this did nothing between the wars to reduce the tenacity of the belief that the market economy was failing, and that it was the process of rationalization (essentially a micro-solution) that offered the way out of the predicament which this posed.

The word 'rationalization' lacked any precise meaning. As the *Economist* remarked: '. . . as often happens with a new immigrant to the language it is undergoing a vogue which has led to its use as a cloak for confused ideas, and sometimes as a badge of respectability for processes of doubtful value'.[10] To some prominent advocates it clearly implied horizontal amalgamation,[11] a view shared by some critics of the rationalization movement who described it, less enthusiastically, as 'a new fangled term to describe the old fashioned device of eliminating competition'.[12] 'I do not much like the word', wrote the economist D. H. Macgregor resignedly in 1934, 'but it has become necessary to use it. It is mainly a question of the scale on which private enterprise should be urged or compelled to reorganize itself by amalgamation.'[13] It is in this sense that the term is used here. However, we shall not lose sight of its wider implications, for the more intelligent apostles of rationalization were careful to stress the interdependence of the various aspects of their programme and the insufficiency of merger by itself as

[10] *Economist* (7 Dec. 1929), p. 1073. For examples of the ragbag of ideas associated with the movement, see: W. T. Davies, *The Rationalization of Industry* (n.d. 1928 ?); L. F. Urwick, *The Meaning of Rationalization* (1929); J. A. Bowie, *Rationalization* (1931); L. J. Barley, *The Riddle of Rationalization* (1932); A. C. Pigou, chairman, 'Problems of rationalization' (discussion), *Economic Journal*, vol. 40 (1930); International Management Institute, *Interim Report on Management Terminology* (Geneva, 1930).

[11] E.g. British Electrical and Allied Manufacturers' Association, *The Electrical Industry in Great Britain* (1929), p. 193; (Macmillan) Committee on Finance and Industry, *Minutes of Evidence* (1931, 2 vols), qq. 3881–3, 7908 (subsequent references to this source are abbreviated to *Macmillan Evidence*).

[12] Attributed to Professor Gregory in W. Meakin, *The New Industrial Revolution* (1928), p. 131, and A. Watson, 'How far can rationalization go ?', *Business* (Aug. 1931), p. 55.

[13] D. H. MacGregor, *Enterprise, Purpose and Profit* (Oxford, 1934) p.v. See also E.A.G. Robinson, *The Structure of Competitive Industry* (1931 ed.), ch. 12 ('Rationalization').

an automatic promoter of efficiency. Lyndall Urwick, for example (perhaps the most consistent and coherent advocate of rationalization), wrote that:

> The mere financial combination of businesses or the wider application of scientific methods of management to existing units of control, can neither of them *by themselves* contribute effectively towards equipping Great Britain with that reorganized national economy which is essential if she is to retain her place among the industrialized nations.[14]

The two were seen as complementary, not as competing, aims.

The call for experimentation in new forms of organization within the firm and for large-scale merger of interests was motivated by the failure of the market economy to produce prosperity and full employment, and as such was a worldwide phenomenon.[15] In Britain, however, perhaps more strongly than elsewhere, the instability and insecurity of these years initiated what was to contemporaries 'a very remarkable change'. Before the First World War,

> . . . while it was admitted that the old theory of competition was not working without disadvantage, it was believed that, all over, these were less than the disadvantages which might result from anything monopolistic. . . . The postwar tendency is to change this attitude.[16]

More complaints of 'overproduction' or 'underconsumption' were heard as one industry after another met problems of overcapacity; and competition was no longer widely accepted as essentially benign, but was increasingly referred to as 'cut throat', 'wasteful', 'unfair', 'destructive' or 'ruinous'. The world faced that 'gigantic paradox' later to be enunciated by Roosevelt but already stated neatly in 1931, by Urwick:

[14] L. Urwick, *Meaning of Rationalization*, p. 134 (present author's italics). For Urwick's other writings on rationalization, see his 'The pure theory of organization with special reference to business enterprise' (British Association, 1930, typescript); *The Management of Tomorrow* (1933); and 'Rationalization', *British Management Review*, vol. 3 (1938). Urwick was the director of the International Management Institute at Geneva, and, on its demise in the 1930s, became a successful management consultant.

[15] L. Urwick, 'The international position', speech at the meeting of the British Association, 1931, reprinted in R. J. Mackay (ed.), *Business and Science* (1931). International Management Institute, *Rationalization and Prosperity*, (Geneva, 1933).

[16] MacGregor, *Enterprise, Purpose and Profit*, pp. 125, 127.

Our control over natural resources is enlarged almost beyond the wildest dreams, even of each preceding decade. The world's capacity for production has been developed to a far greater degree than any corresponding increase in population, especially in the industrialized nations. Yet the peoples of those nations, by millions, are eye to eye with uncertainty, with want, with moral degradation and with despair. We meet under the shadow of the gravest economic crisis which has threatened the material well being of civilization for a century.[17]

The earlier belief in the inevitability of progress in an economy co-ordinated only by the 'invisible hand' of the market looked, in this interwar world, progressively less convincing. One justification on which the earlier faith in a competitive market system had been based was that 'the great complexity of effort necessary to maintain the world's material life cannot be organized, is beyond the control of any form of positive action which humanity can devise....'[18] But this belief crumbled in the face of evidence suggesting that the market also was incapable of that prodigious feat of organization which the world required. The market appeared instead to involve 'seventy-five per cent of our best endeavour ... in unproductive competition, in over-lapping and in confused striving'.[19]

In the place of the market's 'automatic' equilibration, which had once seemed to be its strength, it was suggested that men should sub-stitute a *consciously* and *deliberately* fashioned 'rational' system of industrial regulation. 'There is abundant evidence', Harold Macmillan MP told a management research group in 1934, '... to prove that some form of conscious social direction will have to supplement the old system under which the regulation of our economy was entrusted to the method of trial and error in response to the price indicator',[20] and this sentiment was widely echoed by rationalizers. 'The belief that a more rational control of world economic life through the application of scientific method is possible and desirable'[21] thus grew in influence,

---

[17] Urwick, 'International position', p. 3.   [18] Ibid., p. 22.

[19] Barley, *Riddle of Rationalization*, p. 62.

[20] H. Macmillan, 'The place and functions of large-scale manufacturers in a planned economy', typescript, read to Management Research Group No. 1 (7 Mar. 1934), p. 3. These papers, and many others relating to the early history of research into management, are now in the possession of Mr Harry Ward. I am very grateful to Mr Ward for permission to consult his invaluable archive, subsequent references to which are abbreviated to *Ward Papers*.

[21] Urwick, *Meaning of Rationalization*, p. 27.

amongst men pleased to believe in their creativity and potentiality as cooperators in extending their power of control over economic life by conscious and purposeful organization. Moderate and reflective economists such as Sir Henry Clay were increasingly concerned 'to discourage the hope that the problem, if left to itself, will cure itself, and to argue that the necessary reorganization of the depressed industries will not be affected unless the initiative is taken and the impulse given by some agency outside them'.[22]

The implication of rationality in the term 'rationalization'[23] emphasized that industry could conform to ideas and values whose proponents were growing in confidence and strength in contemporary society, and in particular to the growing awareness of, and faith in, things scientific at the level of popular philosophy.[24] Businessmen and statesmen accepted the common popular theme that advances in science and technology were giving men a growing control over the natural environment and pleaded for a greater recognition that the methods of scientific inquiry could solve social and economic difficulties also. Thus Josiah Stamp, economist, ex-director of ICI, and chairman of the LMS Railway, pleaded for 'a *science* of social adjustment',[25] and Lyndall Urwick in the peroration to his address to the British Association in 1930 challenged his colleagues to accept that:

> It is time for a fresh step forward. Let us use the intellectual conceptions which have given us the great material advances of the machine age, to resolve the new difficulties which that age has created. Let us say as a scientific organization – 'it is intellectually possible; it is in line with our tradition'.[26]

The doctrine of progress through the rational application of scientific principles thus aroused expectations of amelioration in face of evidence of unemployment and instability, strengthening the motive to apply new methods of industry. 'Scientific management' thus attempted, within the rationalization movement, to match the growing success of science in other fields.[27]

---

[22] Quoted, with approval, in 'The case for rationalization', *Economist* (12 Oct. 1929), pp. 652–3.

[23] The etymology of the word is obscure, but the implication of rationality clearly gained its acceptance.

[24] For a parallel, see R. Graves and A. Hodge, *The Long Week-End* (1941), p. 260.   [25] J. Stamp, *The Science of Social Adjustment* (1937).

[26] Urwick, 'Pure theory of organization', p. 10.

[27] E.g. Urwick, *Meaning of Rationalization*, p. 27; Pigou, chairman, 'Problems of rationalization', p. 366; S. Myers, *Business Rationalization* (1932), p. 50.

In this respect, it provided for some businessmen an ideology to
replace the doctrine of competitive free enterprise, whose ethical
foundations and claims to be a gospel of human freedom were being
undermined by socialism, a rival doctrine which in some respects over-
lapped the ideology of rationalization. Competition was disliked by
both socialists and rationalizers, and they both stressed not only
scientific and rational, but also humane, values: 'the glamour of the
perfect, unselfish mechanism hangs about the system of rationalization'.[28] The business classes were, of course, very much aware of the
need to provide a political and economic alternative to socialism. It was
their position that was endangered by depression and it behoved them
to revalidate it by showing that those parts of the market system which
were causing trouble could be excised by the reorganization of private
capitalism. The three-way promise of the rationalizer that 'it is intel-
lectually possible . . . it is materially profitable; it will save our economic
system from disaster'[29] was not easy to ignore. Rationalization readily
became a defensive reaction to the challenge to the existing structure
of power and ownership in industry, which was posed with increasing
urgency by the rapid growth of trade unionism and labour socialism.[30]
It was, then, no accident that in the Mond–Turner talks which marked
the post General Strike rapprochement between moderate employers
and union leaders, the employers' side laid great stress on the potential
social benefits of rationalization.[31]

There was of course an alternative reaction. The inherited faith of
members of the business classes in the 'theology of the market' could be
self-validating. It was not difficult to prescribe a remedy for the deficiencies of the interwar economic system which was consistent with
acceptance of a small enterprise system depending predominantly on
competitive market mechanisms of economic coordination. Blame could
be laid on the trade unions, for example, for introducing undesirable
monopolistic elements into wage determination, or for causing inefficiency through 'ca' canny'. It could be said that government intervention and the expanding public sector had caused a malfunctioning of a

28 'S.R.', 'Advantages and disadvantages of rationalization', *Manchester Guardian Commercial* (18 Oct. 1928).
29 Urwick, 'Pure theory of organization', p. 10.
30 Sir Alfred Mond, 'The rationalization of industry', in his *Industry and
Politics* (1927). Barley, *Riddle of Rationalisation*, ch. 11. Sir William Seager,
'British industry must nationalize or rationalize', *Business* (May 1932),
pp. 9–10.
31 H. Bolitho, *Alfred Mond, First Lord Melchett* (1933), p. 313.

private enterprise system which, had it been left to itself, would have been perfectly healthy. Indeed the very growth of monopolistic combines at the expense of small business could be pictured not as the cure but as the cause of the problem, preventing readjustment of supply and demand by creating price inflexibility. Moreover, all of these things were said, and said frequently, by private businessmen. Yet the flaw in such diagnoses was that, given the political and social situation, the industrial costs of the action they implied were felt by many to be unacceptable. The Geddes 'axe' and the May Report, lock-outs, the forced reduction in miners' wages, the Trade Disputes Act, cuts in local rating for businesses: all these were tried as remedies in this tradition, yet when it came to the point few businessmen viewed with pleasure the prospect of a protracted series of labour confrontations akin to the General Strike.[32] By the same token no government could safely contemplate either cuts in public expenditure larger than those of 1931 or the creation of more widespread distress in already distressed areas. By contrast the replacement of the moral economy of the market by an alternative modified system of economic rationality (which yet preserved something of the privacy of enterprise and the principle of profit) could seem attractive. Conscious action to rationalize industry or to plan the economy, especially when this involved some increase in monopoly power, thus became the vogue in advanced business circles. Many of their goals – profit maximization, the growth of their firms, or merely entrepreneurial exhilaration and empire building – were compatible with and reinforced by the doctrines of rationalization. A programme of merger, inter-firm agreements and 'scientific' management (in short of 'rationalization') thus became the common currency not only of a metropolitan élite of intellectuals (as some of its critics were inclined to imply) but also of businessmen who liked to picture themselves as successful and hard headed.

The critics were correct, however, in pointing to quite virulent opposition to rationalization, an opposition for which in many cases the justification is readily apparent. There was clearly a faddist fringe to the movement, 'distinguished [as one rationalizer admitted] rather by moral enthusiasm than by effective methods',[33] which its critics could

---

[32] On the conciliatory aspects of business opinion, generally, see H. A. Clegg, 'Some consequences of the general strike', *Manchester Statistical Society Transactions* (Jan. 1954), p. 25.

[33] Urwick, *Meaning of Rationalization*, p. 127.

characterize as 'cheap, superficial and popular'.[34] The National Confederation of Employers' Associations (an organization strongly influenced by the many smaller and medium-sized businesses) felt it
necessary to warn the Conservative prime minister that criticisms of the
industrial structure and of businessmen emanated from 'a movement
which seeks to discredit entirely the existing system for the conduct of
industry', and others, too, felt that rationalization came perilously near
to socialism.[35] To this criticism the rationalizers replied in kind,
remarking that their philosophy 'has been met with in many influential
circles by obstinate and destructive prejudice'.[36] The fear that large-
scale organization was not growing as fast as they had hoped led
rationalizers to examine 'the human forces which make for irrational,
illogical, and inefficient organization'.[37] Entrepreneurs were accused of
deriving psychic income from operating a system which was inefficient,
a system which did not maximize profits, or growth. So the debate
raged at a vigorous level of invective and argument, stoked not only by
differing evaluations of the economic efficiency of large firms, but by
radically differing views about the whole future of the enterprise
system. Rival creative images of that future, and not only cost schedules,
were at the heart of this controversy, and this circumstance accounts in
good measure for its virulence.

Economic systems are organized by different societies not only in
response to an objective assessment of the relative costs of alternative
methods of satisfying given wants, but also on ideal grounds – that is,
according to whether a particular economic system will produce as well
as satisfy wants which are considered socially desirable in themselves.
These factors are potentially as important as technical factors in determining the direction of innovation in economic organization, and thus in
inducing movements in the relative costs of market and firm. The
ideologies and self-images of businessmen, which can be studied in
movements of fashion and opinion such as the rationalization movement, can throw light on this process, since rationalizing opinions
would clearly tend to influence the direction of organizational innova-

[34] Mr Roundway, criticizing Professor Florence in Pigou, chairman, 'Problems
of rationalisation', p. 365.
[35] Proposals of the National Confederation of Employers' Associations on the
Depression, dated 10 Feb. 1926, *Baldwin Papers*, vol. 28, pp. 183–95 (the
Baldwin Papers are deposited in the University Library, Cambridge). See also
F. Lee, 'A manufacturer's pointer for British industry', *Ashridge Journal*
(Mar. 1932).
[36] Seager, 'British industry must nationalize or rationalize', pp. 9–10.
[37] P. S. Florence, *The Logic of Industrial Organization* (1933), p. 48.

tions towards management within large firms. Ultimately, however, it would not be the relative debating skills of rationalizers and anti-rationalizers which determined the structure of industry, but rather the objective changes in the relative costs of firm and market which the movement succeeded in inducing. Only if the *ex ante* desire for a more 'rational' control by the firm could be transformed into higher *ex post* net profits based on increased revenues or reduced costs in large firms could it be anticipated that the market would in the long run make its hoped for retreat and yield to the rational brain of the administrator. An interesting parallel case of such a balance between subjective and objective forces can be seen in what seems, at first blush, an unlikely quarter: the evolution of the economic structure of the communist states of eastern Europe. Initially communist governments, on coming to power, have shown a strong preference for direct management and central control. However, the subsequent improvement of material standards has not always been as rapid as was hoped, because of inefficiencies and diseconomies of scale and centralization. As this has been perceived, then, there has been a tendency within the socialist countries to adopt a more decentralized system of production management and a closer approach to market mechanisms. Now, although movements of opinion on economic organization have necessarily been more diffuse and evolutionary than this in Western societies, they nonetheless have had visible effects. It is in this that the major significance of the interwar rationalization movement lies, for it was able to induce investment in innovating techniques of intra-firm organization, and thus motivated the cheapening of management within the firm relative to transactions in the market.

We shall return to the question of investment in the management factor by large-scale firms, but there were clearly other, non-managerial, changes in the relative costs of markets and firms which rationalizers did much to publicize. In particular, competitive market forms of organization required firms to forgo some of the economies of scale which could be achieved by aggregation. There is compelling evidence of the existence of plant economies of scale in a significant number of industries (see Chapter 8), and rationalizers naturally stressed the access to such economies which merger could provide by aggregating the demand schedules of two or more firms in imperfect markets. Technical economies of plant size were not the only significant ones. Contemporaries frequently referred to the marketing economies of merger which were often considered to be the most important economies immediately available,[38] and there were also important financial economies (see pp. 75–6). The view was also gaining ground that only large firms could afford the research necessary in modern science-based industry:

> The small firms find it difficult ... to pay for research laboratories [reported a committee of the Privy Council] ... we believe that some form of combination ... may be found to be essential if the smaller undertakings of this country are to compete effectively with the great trusts and combines of Germany and America.[39]

The experience of foreign countries, where industrialists had reaped (or were supposed to have reaped) the benefits of scale economies, was frequently held out by rationalizers as exemplary; and direct competition from German and American combines, which had already been a source of concern before the war, reinforced this message in the minds of many wartime and postwar merging companies. The house journal of Sperlings, the merger promoters, typifies this mood:

> What has been drilled into us in Great Britain as the result of the general canvassing of our industrial position and prospects in the light of the war, is that there is far too little of 'Big Business' amongst us. While Germany and the United States have been developing huge industrial consolidations, with ample resources, specialized production, collective agencies for sale and distribution, and with a full equipment for scientific research, we in Great Britain have been trying to get along with a multitude of small, rather old-fashioned, manufacturing units, each maintaining its own selling and marketing organizations, not at all alive to science, stubbornly individualistic both in their products and in their attitude towards other firms in the same industry, conscious that the smallness of their installations made for inefficiency and waste and yet debarred from scrapping and rebuilding them on modern lines by the almost prohibitive cost. It is now being generally recognized

---

[38] E.g. A. Marshall, *Industry and Trade* (4th ed. 1923), p. 82; Urwick, *Management of Tomorrow*, p. 70; 'To amalgamate or not to amalgamate', *Business* (Jan. 1928), pp. 14–15.

[39] Committee of the Privy Council for Scientific and Industrial Research, *Report for the Year 1915–16* (Cmd. 8336, 1916), p. 42. See also: Dyestuffs Industry Development Committee, *Third Report* (Cmd. 4191, 1932), p. 8; Management Research Group Minutes (Feb. 1935), p. 20 (*Ward Papers*).

that if we are to hold our own in future we must revolutionize our
scale of doing things; in trade after trade.[40]

There was, throughout the period, a flood of literature praising German
rationalization and American mass production, often contrasting
Britain's industrial structure and performance unfavourably.[41] The
evidence of merger abroad leading to economic expansion and successful competition with English firms in world markets weakened lingering
memories of the great nineteenth century development of English trade
under small-scale competitive conditions. The conventional wisdom
now suggested that, though in the past competition had yielded benefits
for British industry, 'it is also true that a great increase [of production]
took place in the USA and Germany under a condition of restricted
competition and trade organization'.[42] Repeatedly the public rationale
of mergers was the need to catch up with and to 'face on equal terms'
foreign competing firms of larger scale.[43] As Marshall remarked, there
had 'appeared a firm resolve to reconsider British methods in relation
to the problem of the new age; and to the solutions of those problems
which were being worked out in America, Germany and other countries'.[44] As a result of these industrial tendencies, Marshall went on to
argue, 'the supersession of small businesses by large in many industries
is inevitable'.[45]

This tendency was furthered by the more direct intrusion of foreign
influences on the British business scene. A remarkable, and significant,
factor in many of the mergers of this period was the direct and indirect
influence of persons or companies with American or German back-

[40] S. Brooks, 'A British captain of industry', *Sperling's Journal* (Nov. 1921),
p. 19.
[41] E.g. Andrew Stewart, *British and German Industrial Conditions* (1916); Sir
Philip Dawson, *Germany's Industrial Revival* (1926); A. Meakin, *The New
Industrial Revolution* (1928); H. Quigley, 'The large-scale organisation of
production', *Manchester Guardian Commercial* (28 Oct. 1926); D. Warriner,
Combines and Rationalization in Germany (1931); B. Austin and W. F. Lloyd,
The Secret of High Wages (1926); *Macmillan Evidence*, qq. 3883–92, 3895–6,
8326–7, and vol. 2, pp. 147 ff.
[42] 'Report of the Subcommittee on the Advantages and Disadvantages of
Competition', appendix A to Federation of British Industries, *Report of the
Committee on Commercial Efficiency* (privately reprinted with amendments,
1935).
[43] W. J. Reader, *Imperial Chemical Industries. A History*, vol. 1 (1970), p. 455.
British Electrical and Allied Manufacturers' Association, *Combines and Trusts
in the Electrical Industry* (1927).
[44] Marshall, *Industry and Trade*, p. 579.
[45] Ibid., p. 580.

grounds. The recognized expert in large-scale organization in Britain
before the First World War, O. Philippi, was a German, and in the
later rationalization movement leading industrialists such as Sir Hugo
Hirst of GEC, the two Lords Melchett of ICI, and the Renolds of the
Renold & Coventry Chain Company, were all of German origin or
parentage and still strongly influenced by continental models.[46] Large
combines abroad were also aware that they had a technical lead over
their British competitors in large-scale organization and mass production, especially in the new industries. Some of the more enterprising
among them, seeing their competitive advantage, sought to exploit it
directly by entering the British market and establishing subsidiaries in
order to expand their profits in this country. In some industries where
they were unsuccessful they nonetheless frequently spurred their
British competitors into consolidating into larger units themselves.
Already before the First World War the Imperial Tobacco Company
had been formed in such a defensive move against the invasion of the
American Tobacco Company,[47] and such examples soon multiplied.
It was an attempt by the German chemical giant I. G. Farben to extend
its control by acquiring the British Dyestuffs Corporation which preci-
pitated the company's eventual merger with three other British companies to form Imperial Chemical Industries, in imitation of the German
combine.[48] American companies were generally more successful and
thus contributed more directly to the rationalization movement both
by acquiring British companies (which they then merged and expanded)
and also by providing bank finance for rationalization.[49] It was the
American General Electric Company, for example, which acquired and
merged four electrical manufacturers to form Associated Electrical
Industries; and General Motors of the US also acquired Vauxhall
Motors and within a short time joined Ford as a dominant American

---

[46] P. L. Payne, *British Entrepreneurship in the Nineteenth Century* (1974), p. 55.
Sir Hugo Hirst, 'The manufacturer and the state' (address at the annual
dinner of GEC Ltd, 1910). Sir Alfred Mond, *Industry and Politics* (1927).
B. H. Tripp, *Renold Chain, a History of the Company and the Rise of the
Precision Chain Industry* (1956). R. Jones and O. Marriott, *Anatomy of a
Merger, a History of GEC, AEI, and English Electric* (1970), pp. 72–5, 77.

[47] B. W. E. Alford, *W. D. and H. O. Wills and the Development of the UK
Tobacco Industry 1786–1965* (1973), pp. 251–77. See also S. B. Saul, 'The
American Impact on British Industry 1895–1914', *Business History*, vol. 2
(1960).

[48] Reader, *Imperial Chemical Industries*, vol. 1, pp. 439–66.

[49] One banking enterprise which aimed to provide American capital for the
rationalization of British industry was the Finance Corporation of Great
Britain and America, set up in 1928 by the US Chase Bank and ICI jointly.

firm in the British motor industry. Such intrusions from abroad, though frequently resented,[50] were nonetheless powerful reminders to British industrialists of the need to adopt the best practice scale and organization if they were to retain their independence in a period of severe international competition and relatively free movements of capital from America.

The tenor of business opinion had, then, changed greatly since before the First World War, both because of the enthusiastic espousal of the doctrines of rationalization and because of more direct pressure from foreign companies which had espoused them earlier. Of course, few industrialists thought that the creed of rationalization could provide a complete answer to the economic problems of the era, but influential businessmen were increasingly coming to accept that larger scale enterprise could considerably strengthen the international position and the domestic prosperity of British industries. We will later suggest that the impact of this on the waves of merger activity and the rapid increase in industrial concentration in the decade or so after the First World War provide direct evidence of the significance of these movements of opinion (see Chapter 7). Yet already, in anticipation of such statistical confirmation, we can see from more circumstantial evidence that the rationalization movement was likely to exercise a strong influence on the managerial strategies of a significant number of important companies. In a catalogue of contributors to the contemporary debates on rationalization (such as has appeared in the footnotes to this chapter) appear some of the more prominent businessmen of the day. Sir Alfred Mond was involved in the Amalgamated Anthracite and International Nickel mergers; Sir Harry McGowan had presided over the creation of the Nobel combine and was instrumental in merging Joseph Lucas with its major competitors; together, the two men were the architects of ICI. Stamp, Barley and Mitchell were also senior managers in ICI and engaged in merging companies in the chemicals and metals industries. Lyndall Urwick was the founder of a pioneering and successful management consultancy which established itself as a leader in the field of advice on large-scale organization. Sir Mark Webster Jenkinson, the accountant, Reginald McKenna of the Midland Bank, and Dudley Docker, an ex-president of the Federation of British

---

[50] Among leading companies which refused American takeover bids were Huntley & Palmer, GEC, Morris and Austin. A number of firms attempted to change their articles of association in order to bar the possibility of American control.

Industries, gave support to rationalization in engineering, shipbuilding, steel, and chemical manufacture, and they were particularly influential in the mergers with which Vickers were connected. Thus even the opponents of rationalization were constrained to admit that 'the blessed word, with its pseudo-suggestion of the scientific, has already hypnotized quite a number of important people'.[51] The support of 'important people' was in no way fortuitous, for the goal which the rationalizers had set before themselves was no ordinary one: the task was that of revitalizing the capitalist system of enterprise in Britain in what seemed to be its darkest hour.

[51] A. A. Baumann, 'An attack [on rationalization]', *Business* (Mar. 1928).

4

Government: trustbuster
or promoter?

The function of the state is not to preserve
the competitive system.

J. H. JONES, *Social Economics*
(1920), p. 192.

 න්ද

If the rationalizers were to succeed in persuading their fellow businessmen to adopt their methods on a large scale, they would also require
the confidence of public opinion and of the governments of the day.
Yet it was far from clear that such confidence would automatically be
forthcoming, for rationalization could often present its less attractive
public face: that of monopolistic exploitation of the consumer. Businessmen readily admitted that monopolization was an important aim of the
rationalization movement, and, as the *Economist* remarked, 'the en-
hancement of profits by the elimination of competition'[1] was often
uppermost in their minds in planning mergers. There was a justifiable
public suspicion that manufacturers aimed primarily to raise prices or
reduce wages rather than to achieve real economies in production.
The peak of public disquiet had already been reached in the 'Soap
Trust' scandal of 1906, a *cause célèbre* of the prewar amalgamation
movement. William Lever, together with other leading soap makers,
had planned a merger which would give the combined undertaking a
near-monopoly of the British market. Hearing of this, the *Daily Mirror*
and the *Daily Mail* ran a virulent campaign of criticism of the trust
proposals, suggesting *inter alia* that the trust would produce inferior
products at higher prices and provide less employment whilst doing so.
The public outcry and resultant bad will towards Levers and the other
companies immediately reduced the sales of their products and the
proposals for merger had to be hastily abandoned. Trial by press thus
successfully scuppered this early attempt at monopoly.[2]

[1] *Economist* (9 Feb. 1924), p. 241.
[2] C. Wilson, *The History of Unilever*, vol. 1 (1954), pp. 73–88.

It was, however, still possible for the soap firms to merge in a less spectacular fashion, and Lever himself quietly acquired many of his competitors on a piecemeal basis in the two decades following the initial scandal. The earlier debâcle had, however, pointed to the dangers which manufacturers faced when building up a monopoly position, for which the term 'rationalization' could be merely a politic circumlocution. Thus if rationalizers hoped to induce additional large-scale merger activity much would depend on the willingness of the press – and more crucially of the government – to permit businessmen the freedom in these matters to which they felt entitled. There was no shortage of precedents for government intervention to protect the public against monopolies. There was, for example, a long history of government regulation of the railways and similar 'natural monopolies'; and indeed, such regulation had often been requested by business groups themselves.[3] Nor was there a shortage of foreign models on which the government might base its policy. The Sherman Antitrust Act of 1890 in the United States, for example, had been strengthened to include powers to control mergers by the Clayton Act of 1914; and in Germany in 1923 the government had instituted a system of public supervision of monopolies and cartels.

In Britain, however, it was generally the case, at least until after the Second World War, that monopolistic schemes could proceed with little interference from the law. The common law had progressively abandoned its stance against 'restraints of trade' and many judges no doubt shared Lord Chancellor Haldane's view that competition could be undesirable since it 'may, if it is not controlled, drive manufacturers out of business, or lower wages and so cause labour disturbance'.[4] Thus monopolistic practices were freely tolerated and amalgamations could in general proceed without legal intervention. On one occasion, it is true, the aged Mr Justice Eve (a gentleman whose judicial acrimony appears to have descended impartially on all financial bureaucracies)[5] attempted to strike down a merger agreement on the grounds that it was 'a most villainous and mischievous form of finance' which was

---

[3] P. M. Williams, 'Public opinion and the railway rates question in 1886', *English Historical Review*, vol. 67 (1952). Departmental Committee on Railway Agreements and Amalgamations, *Report* (Cmd. 5631, 1911). Sir Alfred Mond, *Questions of Today and Tomorrow* (1912), pp. 141–7.

[4] N. W. Salt *v.* Electrolytic Alkali Co., A.C. 461, 469, quoted in R. B. Stevens and B. S. Yamey, *The Restrictive Practices Court* (1965), p. 31.

[5] Cf. W. J. Reader, *Imperial Chemical Industries: A History*, vol. 1 (1970), p. 447.

'against the public' because it would raise prices.[6] However, though this populist stance had a certain attraction, he was so obviously misguided and ill tempered in his judicial conclusions that they were reversed after only two days by the Court of Appeal:

> ... the court was not concerned to see how the alterations would affect persons outside the company, and, so long as combination was not illegal, it was not for the court to enquire whether the interest of the purchasers would be injuriously affected.[7]

Any serious control of monopolies and mergers between the wars could not, then, depend on the common law; it would require the statutory creation of new powers. The nearest approach to such a government initiative came with the various committees on trusts which deliberated between 1918 and 1921. The first was the Committee on Trusts appointed early in 1918, by the Ministry of Reconstruction, to review the problem of trusts after the war. It noted the expansion of the power of combines both before and during the war, but recommended control by publicity only, a limited prescription which was also endorsed by the Committee on Commercial and Industrial Policy.[8] However, both committees also stressed the benefits of the activities of combinations and were impressed by the need for large-scale organization to meet German and American competition. They also praised the cooperation among manufacturers, for the purposes of standardization and production planning, which had been encouraged by the Ministry of Munitions in the war. This ambivalent approach to mergers and monopolistic arrangements, recognizing the existence of social benefits as well as social costs, was to be present in later discussions of monopoly policy, and, of course, remains today as the central dilemma of antitrust policy.

During the war a system of government price restriction and control of industry had grown up, as a necessary concession to labour in return for their efforts to transform the economy to a war footing. By 1918, however, there was a general feeling in government circles that such controls should be abandoned, typified by Churchill's statement, four days before the armistice, that:

> Our only object is to liberate the forces of industrial enterprise, to release the controls which have been found galling, to divest

[6] Quoted in W. D. Esslemont, 'Some legal aspects of trade combination', *Scottish Law Review*, vol. 44 (1928).   [7] Ibid.

[8] Committee on Trusts, *Report* (Cmd. 9236, 1918). Committee on Commercial and Industrial Policy after the War, *Final Report* (Cmd. 9035, 1918).

ourselves of responsibilities which the state has only accepted in this perilous emergency, and from which, in the overwhelming majority of cases, it had far better keep itself clear.[9]

Although the price controls were a limited safeguard against feared public criticisms of profiteering and monopoly, then, most of them were abandoned between 1919 and 1921.[10] However, in a new initiative, the Profiteering Act of August 1919 made the earning of 'a profit which is, in view of all the circumstances, unreasonable' a punishable offence. The Standing Committee on Trusts, which was set up to administer the new controls,[11] began its meetings in the optimistic belief that it was the harbinger of wider legislation. Its Chairman, Charles McCurdy (formerly Food Controller, and a Coalitionist Liberal MP), reminded its members at their first meeting in October 1919 that 'we are asked to set up machinery for the purposes of starting a general bureau of information on the subject, and ... the government have already started with legislation on these lines which will be made permanent in the Autumn'.[12]

Fifty-seven detailed reports on trusts and prices were to be published in the next two years, but McCurdy's hope of permanency was to prove illusory. As Beveridge judged,[13] the Act was essentially a window-dressing device, and lasted only as long as it was useful in that capacity, becoming a pawn in a political strategy which saw no permanent place for government intervention in industry. With men like Ernest Bevin, J. A. Hobson and Sidney Webb as members, it served to placate the left, but the quality of the membership was a guarantee of nothing more than the provision of information of a usable kind. The Committee was maintained essentially as a safety valve and, given its meagre powers, could provide no coherently articulated antitrust policy. In the Lever Brothers case, for example, it had no powers to reverse the process by which William Lever had built up his dominant position in the soap industry by acquiring competitors. The Committee pronounced not on the merits or otherwise of this process of acquisition but on the demerits of the subsequent price increases. With its initially small powers fast ebbing away, it soon lost all credibility and even lacked the

[9] Quoted, from the *Report of the War Cabinet* (1918), in A. C. Pigou, *Aspects of British Economic History 1918–25* (1947), p. 121.
[10] Ibid., pp. 129–30.
[11] As a subcommittee of the Central Committee established at the Board of Trade under the Profiteering Act.
[12] BT/55/55, *Minutes*, 6 Oct. 1919.
[13] W. H. Beveridge, *British Food Control* (1928), pp. 287–9.

power to require the production of information which it considered relevant. Thus companies involved in mergers and rationalization could, and did, effectively ignore the Committee with impunity.[14]

The history of the Committee on Trusts, though in itself an abrupt historical *cul-de-sac*, is worth recounting because it does help us to understand the basic political determinants of the peculiarly retarded development of monopoly policy in Britain. The political origin of control under the Profiteering Act had been a desire to avoid the social unrest which would follow rapidly rising prices.[15] In April 1920 the Cabinet went on to consider a Trade Monopolies Bill and plans for extending and strengthening government powers in that field. Though the Bill was in fact deferred to a later session (and in the event never became law) the Cabinet debate on the subject is illuminating. It centred round two conflicting political pressures: on the one hand there was 'the widespread expectation of the public that the evils of profiteering would be vigourously [sic] handled' and, on the other, 'the effect of the newly introduced Budget on the business interests represented in the House of Commons and their attitude ... generally, to any undue interference with trade'.[16] The balance of these considerations was calculated to favour acquiescence in labour demands for control of profiteering, for, as they had agreed at a previous meeting, they 'could not afford to take risks with labour. If we did, we should at once create an enemy without our own borders and one which would be better provided with dangerous weapons than Germany.'[17] By early 1921, however, food prices had been falling for some months and the political initiative was shifting away from the interventionist Liberal Coalitionists and their Conservative allies to *laissez-faire* and 'anti-waste' elements. The Profiteering Act was allowed to lapse in May 1921. The protests of the Standing Committee on Trusts[18] and subsequent pleas for the

---

[14] For a review of the weaknesses of the committee, based on its published reports, see J. M. Rees, *Trusts in British Industry* (1922), ch. 11. The surviving minutes of the committee (BT/55/55) confirm the gradual ebbing away of the political support and power of the committee.

[15] *Special Report of the Select Committee on High Prices and Profits, with the Evidence* (Cmd. 166, 1919). PRO/CAB/23/11 (8 August 1919, discussion of Profiteering Bill).

[16] CAB/23/21 (20 Apr. 1920). For an earlier warning of the potential dangers to the public interest inherent in the political representation of business, see T. H. Farrer, *The State in its Relation to Trade* (1902 ed.), p. 123.

[17] CAB/23/15 (quoted in A. Marwick, *Britain in the Century of Total War* (1968), p. 147).

[18] BT/55/55, minutes of the 68th, 69th and 71st meetings (Apr.–May 1921).

machinery to be resurrected[19] fell on the deaf ears of the ministers at the Board of Trade.[20] The Act had been part of the mechanism by which wartime controls were abandoned by a government which had no taste for intervention in the private sector except where decontrol would menace social stability. By 1921 many Conservatives were taking the view that 'propaganda' against profiteering was doing private enterprise more harm than good.[21] Thus, even if the government had itself had the will, it was increasingly undesirable politically to adopt an interventionist line in these relatively new areas of state activity.

The position achieved in 1921 was preserved intact throughout the interwar period.[22] Although the possibility of antitrust laws was considered, the Balfour Committee, which was appointed, *inter alia*, to enquire into 'the present extent of large-scale production, its possibilities and limitations',[23] concluded that 'the case for immediate legislation for the restraint of such abuses as may result from combinations cannot be said to be an urgent one', and pointed to the 'danger that a Bill, even if carefully constructed and safeguarded in its scope, might easily be so changed in Committee as to become . . . a formidable impediment to industrial development'.[24]

This unflattering appraisal of the skills of parliamentary committees in devising suitable legislation to protect the public against the potential dangers of monopoly was placidly received by a predominantly *laissez-faire* Parliament, which acquiesced in this view of its own limitations.

19 E.g. Royal Commission on Food Prices, *First Report* (Cmd. 2390, 1925), paras 342–4.
20 Sir Philip Lloyd-Greame, 139 *H.C. Deb.*, 5s., cols 602–3. He did, however, promise legislation on trusts in a future session, but no legislation was in fact introduced though he remained at the Board of Trade for most of the next decade. The view that his assurance was not honoured because he fell from power with the Lloyd George coalition in October 1922 (see Political and Economic Planning, *Industrial Trade Associations* (1957), p. 19) misses the point. It is Lloyd-Greame's espousal of rationalization, not a change of régime, that explains this policy development.
21 H. G. Williams, *Politics and Economics* (1926), p. 159. Williams later became a junior minister at the Board of Trade.
22 There were residual price surveillance powers in dyes and explosives, and the prices of buildings and food were reviewed periodically, but these initiatives were of little significance.
23 (Balfour) Committee on Industry and Trade, *Final Report* (Cmd. 3282, 1929), 'Memorandum accompanying Terms of Reference', p. iii.
24 Ibid., pp. 191–2. However, the committee did suggest that an investigating tribunal might be valuable, and two minority reports stressed the need for more positive action. Neither of these suggestions was adopted by the government.

Throughout the interwar decades the business lobby, which had been prominent in the Cabinet's calculations on antitrust policy in 1920, remained important.[25] The Federation of British Industries maintained close links with between seventy and eighty MPs, seeking to neutralize criticism in the House of member firms such as J. & P. Coats which occupied monopolistic positions.[26] Industrialists such as Sir Harry McGowan of ICI regularly addressed the 1922 Committee on the virtues of large-scale enterprise, and the tradition of businessmen serving on a Board of Trade advisory committee was firmly established and welcomed by the Cabinet.[27]

In such an atmosphere of 'businessmen's government',[28] *laissez-faire* and capitalism were naturally accepted as the ideal form of economic organization. The contemporary political debate on capitalism versus socialism generated myths and dogmas strengthening the uncritical adherence of the right to a belief in unfettered private enterprise. It was sufficient to express a languid faith in the virtues of the private enterprise system without inquiring too closely whether the competitive conditions necessary for the successful operation of such a system were present. Since socialists and trade unionists had been among the earlier advocates of antitrust and were now pressing for increased government intervention in the market economy, their opponents in government were unlikely to find these policies very palatable, and even a moderate suggestion that the government should sponsor a council to publicize monopolistic pricing could be characterized as a step on the 'slippery slope of socialism'.[29] As a contemporary regretfully recorded, the reaction 'has gone beyond all reasonable bounds, and inhibits the majority of businessmen from exercising any reasonable, intellectual judgement in matters where the action of a government department is conconcerned'.[30] With this convenient configuration of opinion, monopolistic firms could continue to rely on the spectre of state 'interference'

[25] S. Haxey, *Tory MP*, (1939). J. M. McEwen, *Conservative and Unionist MPs 1914–1939* (unpublished PhD thesis, London, 1959), ch. 2.

[26] P. Mathias, *History of the FBI* (typescript, n.d.), pp. 33–4, 44 (available in the library of the Confederation of British Industry).

[27] *Baldwin Papers*, vol. 32, p. 179. Lord Swinton, *I Remember* (n.d. 1948 ?), p. 27.

[28] The phrase is now often used by advocates of a 'new' and efficient government by businessmen. In fact the historical record suggests that the remedy has already been tried! With what success the reader may judge for himself.

[29] T. H. Ryland, minority report, in Royal Commission on Food Prices, *First Report* (Cmd. 2390, 1925), p. 173.

[30] L. F. Urwick, *The Meaning of Rationalization* (1929), p. 124.

with their monopoly position as a remedy which in the long run 'might be worse than the disease'[31] not only for them but for all businessmen.

No other pressure group with an interest in the control of monopoly or mergers really developed in the interwar years sufficiently to have an impact on policy. There was no well-organized consumers' movement, and both the cooperative movement and the Labour Party failed to gain acceptance for their proposals for the investigation of monopolies.[32] The press, though sometimes admitting the logic of a system of public control of monopoly,[33] emphasized also the value of rationalization and the need for a restructuring of industry. Mergers were generally welcomed as a positive sign of industrial vigour, and only in the later 1930s were doubts stirred, as opinion began to turn against schemes of rationalization which showed no regard for the public interest.[34] The general acceptance of the business viewpoint on antitrust was probably strengthened by the high level of unemployment and the declining prices of the time, as official and public opinion reacted with sympathy to the actions of producers trying to improve their lot in admittedly difficult times. 'Sympathies are more with the producer now', commented one economist, 'as with the consumer [formerly]'.[35] Economists, in general, were uneasy about applying their microeconomic work to real economic situations, and, whilst they accepted the economic logic of control, they generally viewed American-style antitrust legislation as ill founded and ineffective.[36] 'Among the intellectual influences on contemporary policy', Robbins sadly noted in 1939, 'there are probably

[31] Committee on Industry and Trade, *Final Report*, p. 189. See also British Electrical and Allied Manufacturers' Association, *Combines and Trusts in the Electrical Industry* (1927), p. 21.

[32] As is witnessed by a failure of the Labour government's Consumers' Council Bill of 1931, which would have created a council of seven to investigate restrictions of competition and have endowed the Board of Trade with price-fixing powers, see *Economist* (4 Apr. 1931), p. 720, 253 *H.C. Deb.*, 5s., cols 2105–200. For the cooperative viewpoint, see *Macmillan Evidence*, qq. 6334–6336.

[33] E.g. *Economist* (7 June 1919), pp. 1040–1; (12 Feb. 1921), p. 273; (26 Mar. 1921), p. 644; (2 May 1936), pp. 248–9.

[34] However, even when strongly condemning trusts, the *Economist* opposed US style antitrust laws with all their 'difficulties and complications'; see 'The cartelisation of England', *Economist* (18 Mar. 1939), p. 552.

[35] D. H. MacGregor, 'Rationalization of Industry', *Economic Journal*, vol. 37 (1927), p. 319. See also *Economist* (22 Dec. 1928), p. 1150.

[36] J. Robinson, *The Economics of Imperfect Competition* (1933), p. 327. A. C. Pigou, *The Economics of Welfare* (1924), pp. 306–13. For a strident, and somewhat inflated, condemnation of economists for their silence on the issue, see W. H. Hutt, *Economists and the Public: A Study of Competition and Opinion* (1936).

few which have been so potent as belief in the inevitability of monopoly',[37] and many politicians also believed, in the words of Philip Snowden, that 'trusts . . . are inevitable. They will continue, whatever obstacles we attempt to put in their path.'[38]

There was, then, for practical purposes, virutally complete acquiescence from government and press in the unfettered movement towards higher industrial concentration. The rationalizing or monopolizing businessman was free to pursue his ends even if the benefits were attained at the cost of the consuming public, and no rationalizer felt it necessary to offer safeguards to the public interest.[39] Furthermore, increasing industrial difficulties, and particularly the high and increasing level of unemployment, created a disposition in both business and government to view the rationalization movement with favour. It was but a short step from this to direct pressure on the government to intervene more positively to promote industrial change both by legislation and by providing finance for mergers and rationalization. There were precedents for such direct action in the government promotion of mergers in the dyestuffs industry during the war,[40] but this policy was implicitly based on the premise that government knew the appropriate structure of industrial firms better than private capitalists and businessmen themselves. Such a doctrine was, of course, a difficult one for the governments we have described, with their essentially *laissez-faire* outlook, to accept. Some government departments, which were already involved directly with industries from which they purchased supplies, did sometimes encourage rationalization by a policy of concentrating orders on a limited number of efficient firms, but such policies were not general.[41] Hence pressure for the government to intervene in areas where it had no such direct interest mounted. Some of the younger Conservative MPs, and the government's own economic advisers, for example, pressed the view that there was scope for greater state intervention to

---

[37] L. Robbins, *The Economic Basis of Class Conflict* (1939), p. 45.

[38] Quoted, with approval, in R. Boothby *et al.*, *Industry and the State: A Conservative View* (1927), p. 47.

[39] Cf. W. Letwin, 'The past and future of the American businessman', *Daedalus*, vol. 97 (1969).

[40] Reader, *Imperial Chemical Industries*, vol. 1, pp. 258–81.

[41] E.g. the Post Office speeded the rationalization of the cable industry in this way; see Monopolies and Restrictive Practices Commission, *Report on the Supply of Insulated Electric Wires and Cables* (1952), p. 40. But, for the failure of government to adopt such a policy elsewhere, cf. P. Fearon, 'The British airframe industry and the state, 1918–35', *Economic History Review*, vol. 27 (1974).

change the structure of the more traditional industries.[42] At least one
senior Conservative economic minister felt not only that more mergers
and rationalization were needed, but also that industrial management
was incapable, by itself, of solving the problems which they posed,[43] a
view which was shared by some senior industrial managers with
government contacts, who doubted the capacity of their colleagues in
other firms.[44] Such criticisms could not, however, be publicly voiced by
senior ministers for fear of alienating an important section of the
Conservative Party's political support.[45] Instead the government
tended to look hopefully to industrialists for evidence that they were
hatching appropriate schemes, and industrialists glowingly obliged
by pointing to their proven zeal for merger and cooperative
schemes.[46]

Mergers were thus in general welcomed by government ministers,
who (in the words of the President of the Board of Trade) saw 'before
themselves the modest task of clearing away difficulties and of giving
the necessary support and shelter to private enterprise to cope with its
problems on sound principles'.[47] The only real concession to the
rationalization lobby was to be not the wholesale promotion of merger
but the removal of minor disincentives. In the 1927 budget the payment
of stamp duties in cases where a new company was formed to merge
two or more existing companies was remitted, a concession which was
generally considered to accord with common sense and equity.[48] This
relatively modest approach was supported by industrialists, many of
whom were ardent rationalizers but felt that bankers rather than the
government were the appropriate agency to bring pressure to bear on

[42] Boothby et al., Industry and the State. Economic Advisory Council, Report
on the Cotton Industry (Cmd. 3615, 1930).
[43] Sir Arthur Steel-Maitland, Minister of Labour 1924–9; see Baldwin Papers,
vol. 30, pp. 9–17.
[44] W. H. Coates, 'Memorandum', dated 5 December 1933, criticizing a National
Confederation of Employers' Organizations report because it had 'assumed
that the economic factor of management and organization in British industry
needs no comment' (ICI archives).
[45] Baldwin Papers, vol. 30, pp. 41–3 (memorandum dated Feb. 1929).
[46] E.g. letter from Sir W. J. Larke to Sir P. Cunliffe-Lister, dated 3 Dec. 1925
(Baldwin Papers, vol. 27, p. 232).
[47] BT/55/49, Sir P. Cunliffe-Lister, memorandum of 1927, p. 18.
[48] Finance Act, 1927, section 55. P. F. Simonson, The Law Relating to the
Reconstruction and Amalgamation of Joint Stock Companies (3rd ed. 1919),
ch. 5. A. Mond, 'National Savings, profits and double taxation', in his
Industry and Politics (1927). The concession was very rigidly circumscribed
and some mergers still failed to qualify; see R. W. Moon, Amalgamations and
Takeover Bids (2nd ed. 1960), ch. 4.

recalcitrant firms which would benefit from merger.[49] They often discussed their own larger mergers with ministers, both informally and through official channels,[50] but many businessmen feared accepting state money 'because they realized that government capital would be bound sooner or later to lead to government control'.[51] In fact, however, even when less timid (or more impecunious) businessmen did make direct application to the government for financial assistance for schemes of rationalization, they were invariably refused, on orthodox Treasury grounds.[52]

However, the political pressure for a more forward policy on rationalization was leading somewhere, even if only in an indirect way. Montagu Norman at the Bank of England viewed it with increasing alarm and was particularly anxious to head off any attempt to renew the Trade Facilities Act for the purpose of providing government support for mergers.[53] He hastened to assure the government that adequate finance would be available privately for worthwhile amalgamation schemes in the basic industries, and began talks with the banks to canvass this object. The assurances fell on willing ears, and the government was grateful to Norman for taking the burden of forcing mergers on recalcitrant industrialists in the steel and cotton industries out of the government's hands and putting it into the hands of the banks. Churchill, then Chancellor, used Norman's assurances in his 1929 budget speech to nip criticism in the bud,[54] and his Labour successors, Snowden and Thomas, were able to develop this approach when Norman announced the formation, by a consortium of banks with Bank of England backing, of the Bankers' Industrial Development Company (see pp. 73–5).

The Labour government of 1929 to 1931 did not, of course, entirely

---

[49] *Macmillan Evidence*, q. 4111. A. F. Lucas, *Industrial Reconstruction and the Control of Competition* (1937), p. 139.

[50] Lord Chandos, *Memoirs* (1962), p. 125. Reader, *Imperial Chemical Industries*, vol. 1, pp. 251, 247. A. G. Whyte, *Forty Years of Electrical Progress* (1930), p. 93.

[51] *Economist* (16 Jan. 1932), p. 107.

[52] E.g., for Churchill's refusal to finance the Vickers-Armstrong merger, see J. D. Scott, *Vickers: A History* (1962), p. 165. The merger proceeded nonetheless.

[53] A file of J. H. Thomas, who inherited this policy in 1929 (BT/56/14, CIA/621), states that his predecessors 'had in mind an adaptation of the Trade Facilities Acts to stimulate amalgamation and reorganization'. It cannot have advanced very far. The Trade Facilities Act lapsed in 1927. For examples of the opposition to renewal, see *Baldwin Papers*, vol. 30, p. 18. And on Norman's views see Sir Henry Clay, *Lord Norman* (1957), ch. 8.

[54] 227 *H.C.Deb.*, 5s., col. 58.

share the Conservatives' ideological aversion to the belief that government decisions on industrial matters could be superior to private ones, yet Labour ministers, many of them with deeply conservative instincts on economic matters, went little further than their predecessors. Many Labour leaders, and their advisers, sympathized with the broader views of the rationalization movement,[55] but they had doubts about its implications for monopoly power and industrial unemployment. Successful rationalization could, perhaps, through technical and organizational innovation, release economic resources – capital, labour, management – from redundant uses and deploy them to new, higher productivity uses. The release of labour was, however, a mixed blessing for a government beset by social distress and cries for political relief from near-derelict communities which already had vast labour surpluses. Although some rationalizers argued that cost reductions achieved by merger could increase aggregate demand (given a competitive pricing response, income effects, and high price elasticities of demand), most economists had grave doubts about the ability of mergers through rationalization to solve the fundamental problem of mass unemployment.[56] Thus the radical suggestion that the Board of Trade should take over BIDC's functions and play a more effective role in merger promotion was rejected.[57] Yet there was still some pressure on the government to consider a more active role in industrial reorganization, and to meet this Sir Horace Wilson, acting as Chief Industrial Adviser, was charged with monitoring the progress of rationalization in various industries. His brief was to stimulate discussions between industrialists in which he was to attempt, by the use of moral suasion, to induce them to merge.[58] However, without the financial powers which gave the later Industrial Reorganization Corporation its leverage, his initiatives came to little; and, given the government's lack of a parliamentary majority, it was obvious, as the *Economist* pointed out, that hints of compulsory amalgamation if voluntarism failed were mere empty threats.[59]

The fall of the Labour government in 1931, and the world economic

[55] *Macmillan Evidence*, qq. 7961–6 (Ernest Bevin). G. D. H. Cole (ed.), *Studies in Capitalism and Investment* (1935), p. 42. *Economist* (23 May 1931), p. 1090.

[56] T. E. Gregory, 'Rationalization and unemployment', *Economic Journal*, vol. 40 (1930). J. A. Hobson, *Rationalization and Unemployment* (1930). *Macmillan Evidence*, qq. 4886, 4951–7, 5003. But cf. L. F. Urwick, 'Rationalization does *not* lead to unemployment: a reply to Professor Gregory' (typescript in *Ward Papers*, 1930).

[57] BT/56/37, Board of Trade comments on the Mosley Memorandum.

[58] BT/56/14. BT/56/37. BT/56/43.

[59] *Economist* (25 Apr. 1931), p. 885.

crisis of that year, brought the adoption of the 'great policy' of protection in 1932. Tariffs were important for the Conservatives in the National government in their effect on the psychology of interventionism, for, through the Import Duties Advisory Committee, more positive government pressure on industry was legitimized. In the steel industry, in particular, the IDAC was able, through moral suasion and the threat of reducing the level of tariff protection, to encourage restrictive agreements and mergers. However, the approach remained a cautious one and the more adventurous schemes of reorganization were shelved as the committee trod the new ground of government usurpation of entrepeneurial decision making powers warily.[60] Whilst the IDAC certainly played a part in 'breaking down individualism and educating businessmen in many industries to work together',[61] most of its ideas for encouraging the restructuring of industry do not appear to have been accepted. In the Lancashire cotton velvet industry, for example, the IDAC, despite its threat of withdrawal of duty, was no more successful in persuading industrialists to agree to mergers than Sir Horace Wilson had been in 1931.[62]

Though the National government maintained the policy of talks with industrialists which had been initiated by its Labour predecessors[63] and extended this policy through the IDAC, it continued to eschew undue dictation to private enterprise. Walter Runciman, the new President of the Board of Trade, and Leslie Hore-Belisha, his parliamentary secretary, were not enthusiasts for government intervention and Runciman was sincere when he told the British Bankers' Association in 1934 that he would resist the pressure on him for more intervention, adding that:

We may easily exaggerate the importance of mere size, and there are not wanting signs that in some quarters the limits to size have already been reached. . . . The truth is that in these days we have rather underestimated the value of competition. . . . It is in the

---

[60] Lucas, *Industrial Reconstruction*, p. 331. Sir Herbert Hutchinson, *Tariff Making and Industrial Reconstruction* (1965).

[61] Hutchinson, *Tariff Making and Industrial Reconstruction*, p. 78.

[62] Ibid., pp. 125–30. They did, however, by threatening withdrawal of duty gain the adoption of a price agreement.

[63] Sir Horace Wilson (interview, 28 July 1969) confirmed that he and Sir Leonard Brewitt continued to visit industrialists to draw their attention to the facilities offered by banks, and liaised weekly with Charles Bruce-Gardner of the BIDC.

preservation of competition ... that we can ... preserve the prosperity of our great commercial and industrial organizations.[64]

The views of interventionist businessmen like Harold Macmillan
may have harmonized with the collectivist streak in Tory philosophy,[65]
but their influence on government policy between the wars was minimal.

Essentially, then, the position of interwar governments was to
eschew not only the role of trustbuster but also that of trust promoter.[66]
Both of these policies are, we have argued, explicable in terms of the
ideology and motivations of the governments of the day. They were for
the most part more certain of their commitment to a capitalist system
of enterprise, privately owned and as far as possible free of state
intervention, than to the ideal of competition or the goal of planning
through large-scale enterprise. When Stanley Baldwin told his Bewdley
constituents at Worcester in 1929 that 'no one rejoices more than I do
to see these industrial problems taken directly out of the hands of
politicians, who have never been fit to handle them',[67] it was not mere
political rhetoric but a genuine statement of his personal preference,
and (making an allowance for the false modesty appropriate to such
gatherings) an important strain of Conservative thinking. Many of his
ministers shared this view and many of their supporters felt that it best
served the interests of private capital. In this atmosphere, the ideal of
laissez-faire – a laissez-faire that meant freedom from enforced merger
as well as freedom from enforced competition – was still a powerful
factor. If ministers were being forced by circumstances into a more
forward interventionist policy in some spheres, many of them remained
determined to limit it, and, in the case of the promotion of mergers and
large-scale enterprises in manufacturing industry, they were largely
successful.

Now these conclusions may appear to conflict with the conventional
interpretation of the period as one of willing, extensive and increasing
government intervention in the economy. Partly this interpretation arose
because contemporaries were extremely sensitive to the extension of the
powers of the state, and in many spheres the state was playing a larger

[64] Speech (9 May 1934) reported in *The Times* (10 May 1934), p. 16.
[65] S. H. Beer, *Modern British Politics* (1965), p. 293–7.
[66] There was only one case of a legislatively backed merger in manufacturing
industry in the 1930s: the unification of six beet sugar processing companies
to form the British Sugar Corporation. The government was heavily involved
financially in subsidies to beet growers, and this motivated the exceptional
intervention of the Sugar Industry (Reorganization) Act, 1936.
[67] Reported in the *Financial News* (7 Jan. 1929), p. 7.

role than it had before the war. It was thus a short but plausible step for free marketeers who disliked monopoly to assert that it arose because of political meddling rather than from tendencies inherent in the competitive system itself.[68] Even moderate organs of opinion such as the *Economist*, which recognized the undesirability of the extreme *laissez-faire* position in the conditions of modern industry, became convinced by the late 1930s that in many fields the government had gone too far.[69] In later years, then, it was natural for the government itself to be branded as the cause of the ills of increased industrial concentration. Some later antitrust campaigners believed, quite incorrectly, that the government had been a pliant tool in the hands of manufacturers seeking legislative sanctions for mergers.[70] This approach was also strengthened by the interpretative contribution of Marxist historians, who, wishing to establish the mutual support of private capital and the bourgeois state, have presented increased concentration as 'almost invariably fostered by a benevolent government'.[71]

The purpose of this chapter is not to deny the correctness of these views on government action over much of the economy. Had we been concerned with policy towards the electricity industry, the railways, mining or agriculture, our conclusions might well have coincided, for there was an increasing commitment of government finance and legislative support to these industries, whose special economic characteristics of strategic importance dictated government involvement. In the case of manufacturing industry, however, it was sometimes admitted by even the most ardent free marketeers that they had little evidence of direct government involvement;[72] and the *Economist*, while criticizing proposals for legislative support of monopolistic cartels, even argued that stronger measures to promote large firms in manufacturing by compulsory amalgamation were required.[73] Thus, whilst it is true that the governments of the day, and a large section of public opinion, gave their general approval to the contemporary trend towards merger and large-scale enterprise, actual policy in this sphere was remarkably non-

---

[68] L. Robbins, *The Great Depression* (1934); and his *Economic Basis of Class Conflict*, pp. 50–1.
[69] Compare the adverse comments on Baldwin's extreme *laissez-faire* in 1929 (*Economist*, 12 Jan. 1929, p. 47) with the articles on 'The new feudalism' (*Economist*, 2 Apr. 1938, pp. 2–3) and 'The cartelisation of England' (*Economist*, 8 Mar. 1939, p. 552).
[70] E.g. N. E. H. Davenport, *Vested Interests or Common Pool* (1942), p. 54.
[71] E. J. Hobsbawm, *Industry and Empire* (1968), p. 181.
[72] Robbins, *Economic Basis of Class Conflict*, pp. 59, 62.
[73] *Economist* (8 March 1939), p. 552.

committal throughout the period.[74] One further reason why governments were happy to maintain this stance was that, even without the stimulus of government involvement, there was a substantial increase in merger activity and industrial concentration in these years (see Chapter 7). This movement resulted from the decisions of private and corporate entrepreneurs to expand, decisions taken principally in response to the signals of the market and the quest for private profit, rather than at government behest. Governments in general would give their encouragement and their blessing to these decisions (and this was welcomed by businessmen), but they would give little else. In particular, they made it clear that the financial and managerial resources which the new larger scale enterprises required had to come from private financial sources and from the businessmen themselves. Ministers often opened their hearts and praised mergers, but the Treasury purse was firmly closed.

[74] However, for the government's more positive support of restrictive business practices, see pp. 154–6 below.

5

Capitalist ownership and the stock market

The divorce between responsibility and ownership worked out by the
growth and development of Joint Stock Companies . . . provide[s]
one of the clues to the future. Private enterprise has been
trying . . . to solve for itself the essential problem
. . . [of] how to establish an efficient system of
production in which management and responsibility are in different hands from those
which provide the capital, run the
risk, and reap the profit.
Liberal Industrial Inquiry, *Britain's
Industrial Future* (1928), p. 100.

છ૭

In the nineteenth century the extent of the personal wealth and credit
of the owning family or partners usually established an upper limit to
the size of a firm. However, this constraint did not rule out large firms,
for the power of compound interest in building up a single family's
fortune over the years was considerable. Family firms which had
started business with assets worth only a few hundred pounds could,
over two generations or so of successful business, become millionaire
firms; and the capacity of such firms to borrow funds for further expansion also expanded proportionately to their assets. Those few owners
who became millionaires several times over – like the old established
brewing families or the Wills family in the tobacco industry – were able
to finance from their own resources companies which were among the
largest in existence at the turn of the century. However, the number of
families amassing such large fortunes was inevitably limited, and
small and medium-sized enterprises remained more typical of manufacturing industry as a whole. These groups of wealthy, and not-so-wealthy,
capitalist owner-entrepreneurs together held beneficial ownership of a
large proportion of the national capital stock. Some wealthy families,
it is true, had liquidated their interests and spread the ownership of
their firms more widely through the medium of a public flotation on
the stock exchanges, but the majority of British manufacturing firms

were still typically owned and controlled by single families or by partnerships. It is, then, hardly surprising that in 1911 the wealthiest 10 per
cent of the adult population owned as much as 92 per cent of the total
personal wealth in England and Wales.[1]

The further growth of large-scale enterprise, which the rationalizers
were urging, would clearly require a modification of this régime of
private capitalist ownership, since larger firms would in general imply
the creation of industrial units, the ownership of which would effectively
be beyond the financial resources of any single family. There were a
number of possible ways of raising the additional finance. One of these
was the recruitment of capital with legislative support from the state.
This source had been used extensively by public utility enterprises,
and one contemporary estimate suggested that by 1928 two-thirds of
the capital required for large-scale undertakings in the economy as a
whole had been raised with state assistance of various kinds.[2] In manufacturing industry, however, as we have seen, there was a firm political
determination that the finance should come from private sources.
Another option open to businessmen, and one which in the abstract
would have gained the approval of many capitalists, was the ploughback
of internally generated profits into firms, thus enabling them to grow
in the traditional manner of the family firm; the owners would thus be
able to finance their own firms throughout if growth could be spread
over an appropriately long period. There were many large firms between
the wars which were able to rely principally on this option, including
sizable old established family firms such as Pilkington in the glass
industry, and new, rapidly expanding enterprises such as that of
William Morris in the motor car industry. Morris, who began as the
owner of little more than a bicycle shop, was able to expand on the
secure financial basis of very high levels of profits (from his innovation
of mass production in the car industry at Cowley) and retained personal
ownership and control, until in 1935 he floated Morris Motors ordinary
shares on the London stock exchange. He had earlier refused a merger
with both Austin and Vauxhall because his large private fortune would

---

[1] J. Revell, 'Changes in the social distribution of property in Britain during
the twentieth century', *Transactions of the Third International Congress of
Economic History*, vol. 1 (Munich, 1965), p. 379. Of course this figure reflects
the concentration of other forms of asset holdings, and particularly of landed
property, as well as of manufacturing wealth.

[2] Liberal Industrial Inquiry, *Britain's Industrial Future*, p. 74. This figure
includes, for example, capital raised by privately owned electricity undertakings possessing a statutory monopoly.

have been insufficient to gain full control of their capital, but even without such mergers, he gained a leading position in the British car industry, and was able to achieve many of the benefits of large-scale production to which the rationalization movement had drawn attention.[3]

Yet this pattern of growth could hardly have been widely adopted, for in other industries levels of profits and the capacity for internal growth were rarely as high as in the motor car industry. Even if it had been possible in economic terms, the political consequences of the further concentration of an (already highly unequal) distribution of wealth in the hands of fewer owners would hardly have been attractive. Yet higher concentration in industry could not have been attained under a régime of direct capitalist ownership without such a shift towards still greater inequality in the ownership of manufacturing wealth. The way out of the dilemma posed by such financial and political constraints was provided by the further development of the stock market both as a provider of new funds and as a market in titles to existing industrial assets. Industrialists were able, through the stock market, to recruit capital in large amounts and from many sources. Provided that the prospects of large-scale enterprise were sufficiently attractive to call forth such investment, the capital resources of a number of moderately wealthy individuals could in principle, through the medium of the stock market, be aggregated to provide enough finance for the largest of firms. It was on this pattern, rather than in the nineteenth century mould, that the financing of large enterprise was to be achieved.

This tendency was, of course, already exemplified in the quoted manufacturing companies of the late nineteenth century, but it was only from the 1920s that the consequences came to be widely remarked.[4] One widely quoted survey of a sample of the shareholdings in large companies in the interwar period, for example, showed that the average holding of capital was only £301.[5] There were still, of course, many quoted companies with less dispersed ownership, and some in which the original founding families held substantial shareholdings and were still able to secure representation on the board and effective managerial control. The transition to a more modern pattern was thus a gradual one, but many quoted companies soon had boards of directors who no

[3] P. W. S. Andrews and E. Brunner, *The Life of Lord Nuffield* (Oxford, 1955).
[4] E.g. R. H. Tawney, *The Acquisitive Society* (1921).
[5] (Balfour) Committee on Industry and Trade, *Factors in Industrial and Commercial Efficiency* (1927), p. 128.

longer held their positions by virtue of being the largest shareholders.[6]
This trend has continued, and many of the largest firms of today – firms
like ICI and Unilever with market capitalizations of over £1000
million – would effectively be beyond the personal wealth of any single
individual.[7] The dispersion of ownership among many shareholders
has thus become not merely a convenience but even a necessary condition of their existence.

This change has sometimes been dubbed the 'democratization of
ownership', but it was not in fact accompanied by a very significant
shift in the overall social distribution of property ownership. In 1960,
for example, the wealthiest 10 per cent of adults still held 83 per cent of
the personal wealth in England and Wales, and the vast majority of
families still had no share in the ownership of industrial companies.[8]
What had happened over the period since 1911 was not that a significantly larger proportion of families had become large shareholders, but
rather that the minority of wealthy families no longer held their wealth
in single companies in which they were also directors, choosing instead
to spread their wealth over a wider range of assets. It is the divorce of
ownership and control, rather than the democratization of wealth, that
has characterized the twentieth century development of capitalist
enterprise. Capitalists have changed their own patterns of asset holding,
and the distinction between the *rentier* investors who provide capital
and the directors and managers who organize production has in the
process become more clearly articulated; but the general distribution of
wealth in the community has remained remarkably stable.[9]

The motives which have led wealth holders to exchange a position
of owning and controlling moderately sized enterprises for one of

[6] P. Sargant Florence, *Ownership Control and Success of Large Companies
1936–51* (1961). W. H. Coates, 'Administration and capital', *British Management Review*, vol. 3 (1938), p. 62. 'Shareholders and control', *Economist*
(30 Mar. 1929), p. 691.

[7] In 1968–9 the chairmen of the largest 100 industrial companies in Britain
on average controlled only 2½ per cent of their companies' equity and the
whole boards of directors only 7½ per cent; see A. Lumsden, 'Wealth and
power in Britain's top boardrooms', *The Times* (9 Sept. 1969).

[8] Revell, 'Changes in the social distribution of property', p. 379. It could,
however, more plausibly be argued that, because of the nationalization of
some industries and the spread to less wealthy groups of insurance and
pension funds, the beneficial ownership rights of industrial capital are now
spread more widely.

[9] Of course, some fortunes had been made and others had been lost, so that
neither the wealthy families of 1960, nor the assets which made up their
wealth, were the same as those of 1911. However, inheritance has ensured
that there has been remarkable stability in this sense also.

merely owning a diversified portfolio of shares in quoted companies[10] are many and varied. What seems to have happened in the case of the majority of family firms in manufacturing industry is that the family owners have at some stage taken the decision to sell out to an already quoted company, or, alternatively, to make a direct flotation of their company, possibly together with others, on the new issue market. The proceeds of these transactions have then typically been reinvested in a range of stock exchange and other securities, diversified over a number of industrial and commercial fields. We have already seen that the partial liquidation of family interests through company flotations became increasingly popular from the late nineteenth century onwards.[11] One possible explanation of this is that death duties, introduced in 1894, made it increasingly difficult to pass on total financial control of the family firm from one generation to another. However, estate duties were notoriously easy to avoid, and the Colwyn Committee, which considered the evidence, concluded that there were very few estates which did not have sufficient non-business assets to cover the cost of death duties incurred.[12] In cases where a business was sold on the death of the owner, then, this was more likely to be because of the wish of the family to diversify their holdings, or possibly to secure management succession, than because of inheritance problems.[13] Moreover, it is also clear that the vast majority of businesses which passed from the private to the publicly quoted sector did so not as a result of the death of the owner but rather on his explicit, considered and conscious choice. Whether we are thinking of inherited family firms or of relatively new enterprises (such as had been built up by Morris), owners voluntarily supplied a large flow of business assets to the expanding quoted sector, and death duties played little part in their motivation.

They were led to do so, we may safely conclude, because the private

10 This is an oversimplified view of what actually happened: the holdings of wealthy families were often already diversified in the nineteenth century and their later portfolios often included a range of investments other than quoted companies. However, contemporary wills and the private financial papers of businessmen indicate that a diversifying movement of this kind was occurring within such broader movements.

11 See Chapter 2. Of course, families could also liquidate their holdings more gradually by reinvesting their profits in stock market securities (rather than in the family firm itself, as they had in the past). The effect is similar.

12 (Colwyn) Committee on National Debt and Taxation, *Report* (Cmd. 2800, 1927), appendix 20.

13 J. R. Allan (ed.), *Crombies of Grandholm and Cothal 1805–1960* (Aberdeen, 1961), p. 122. B. Newman, *100 Years of a Good Company* (1957), pp. 95–7. *Macmillan Evidence*, q. 3704.

advantages to themselves of the new larger companies which could be created through the medium of the stock exchange were high. The gains in efficiency, to which the rationalization movement was drawing attention, were potentially large, and, in so far as these could be captured as returns to capital, they were attractive to capitalists. The returns from economies of scale and monopoly powers would accrue to wealth owners only if they were prepared to consolidate their assets into larger units through the medium of efficient quoted companies, and this in itself frequently required a divorce of ownership and control. But there were also special considerations which enhanced the private profitability of switching from direct ownership to portfolio investments. Partly this was a question of tax avoidance. Higher rate tax payers – and most owners of family firms in manufacturing industry would be in the supertax bracket – generally preferred capital gains to income, since in Britain until 1965 most capital gains were free of tax. If they were able to capitalize the future income of their business, by selling out, the controlling family could thus frequently reduce their tax liability. The transaction could be profitable also to the new owners since, if they were paying a lower rate of tax, the effective rate of return on the future income flow would be higher: thus the transaction could be advantageous in these cases to both buyer and seller.[14]

A further advantage for family owners in selling out arose from the volatile conditions of the contemporary stock market. Keynes's unflattering comparison of the stock market to the casino[15] had its justification in the erratic speculative movements of share prices. The increased marketability of shares, coupled with serious informational imperfections in the capital market, laid the way clearly open for speculative activity.[16] For those who bought shares when the market was high only to be caught by a stock exchange collapse, this volatility would hardly endear them to the holding of assets in a general share

14 The 1922 Finance Act closed a loophole which had also allowed *past* profits to be retained by closely held companies and subsequently capitalized by sale, but the capitalization of future profits by sale remained possible; see L. H. Seltzer, *The Nature and Tax Treatment of Capital Gains and Losses* (New York, 1951), p. 260. Cf. J. K. Butters, J. Lintner and W. L. Cary, *Effects of Taxation: Corporate Mergers* (Boston, 1951); R. Lacks, 'Income tax on capital profits', *Modern Law Review*, vol. 6 (1943); (Cohen) Committee on Company Law Reform, *Minutes of Evidence* (1943), p. 126.
15 J. M. Keynes, *The General Theory of Employment, Interest and Money* (1936), p. 159.
16 N. J. Grieser, *The British Investor and his Sources of Information* (MSc Thesis, London, 1940). Departmental Committee on Share Pushing, *Report* (Cmd. 5539, 1937).

portfolio. There is, however, compelling evidence that the majority of private companies sold out their business interests by more advantageous manipulations of the stock market. They were helped in this, as they had been before the First World War (see pp. 21–2), by company promoters who provided investors with 'advice'. These men persuaded industrialists to merge and publicly float their firms, ostensibly to gain access to economies of scale, monopoly power and other benefits. However, the promoters' profits were derived from the *ex ante* expectations of these benefits rather than from their *ex post* realization. Many of them therefore attempted to inflate the value of expected gains from postmerger integrations and to capitalize these in public issue at inflated prices which reflected these exaggerated claims, thus enhancing their own profits.[17]

The speculative booms were intensified by this activity, but this did create a clear profit opportunity to the important sector of owner-managed 'family' firms. It seems probable that the 'shadow' valuations of shares by owner-managers and other holders of stock in unquoted companies are in general less volatile than the valuations placed on titles to firms' assets by the stock market. This valuation discrepancy is positive in a boom and creates a margin of profit in a flotation for both the original owners and the promoters. It was thus in the fevered stages of stock exchange upswings that promoters were most active in their business of creating and exploiting speculative hopes by playing on the acquisitive instincts of investors and exaggerating the values of shares. There was a close, positive correlation between the sales of firms by their private owners and the level and rate of change of share prices. Contemporaries noticed that new issues to acquire existing assets were highest in stock market upswings and declined both absolutely and relatively to other new issues when the market slumped.[18] Further, it can also be shown that the level of merger activity (much of which consisted of unquoted firms being acquired by quoted ones, or merging with others in order to make a public issue economically) was positively correlated with share prices throughout the interwar years.[19]

[17] H. O. O'Hagan, *Leaves from My Life* (1929). F. Lavington, *The English Capital Market* (1921), pp. 213–14. A. Marshall, *Industry and Trade* (4th ed. 1923), pp. 330–4.

[18] G. D. H. Cole, *Studies in Capital and Investment* (1935), pp. 124–5. But cf. A. T. K. Grant, *A Study of the Postwar Capital Market* (1937), pp. 130–2, 159–61; R. F. Henderson, *The New Issue Market* (1951), pp. 24–6.

[19] This can be seen by comparing the merger statistics in Appendix 1, below, with the share price index. For a fuller econometric study of the relationship, see L. Hannah, *The Political Economy of Merger in British Manufacturing*

The existence of high profits in the mediation of the supply of firms and the demand of investors for securities, by the promotion of new issues and merger flotations, is also well documented.[20] Both the post-war boom of 1919–20 and the rampant speculation of the later 1920s induced stockbrokers, issuing houses and *ad hoc* promoting syndicates to feed the speculative fever with dubious issues. 'Probably never in the history of modern trade and industry', suggested the *Economist*, 'was the net spread by the company promoter as it is today.'[21] Among them were many who could restructure industry to establish the basis of profitability which the high market capitalizations implied, and even promoters as unscrupulous as Clarence Hatry appear to have had a genuine faith in rationalization and some successful and logical mergers to their credit.[22] However, large profits could also go to the exaggerators and deceivers, and mergers offered a particularly fertile ground for such men, who could:

... realize the advantages that arose from manipulating, not a single company, but a group of companies. A 'parent company' with a number of subsidiaries at its command could do much to baffle the public as to its true state of prosperity. Not only could loans to, and investments in, subsidiaries be so manipulated that balance sheets concealed the true position of the whole group, but sales of rights could be so arranged that a subsidiary could show and transfer a dividend to the parent company which to the average investor may not have been indistinguishable [*sic.* distinguishable?] from a true trading profit.[23]

Thus developments in the stock market not only failed to improve the flow of information to investors but positively tended to distort it, and thus encouraged mergers of dubious economic logic.

Industry between the Wars (unpublished DPhil thesis, Oxford, 1972), pp. 179–186.
[20] E.g. E. V. Morgan and W. A. Thomas, *The Stock Exchange* (1962), p. 106; E. T. Hooley, *Hooley's Confessions* (n.d. 1925?), pp. 303–6; E. V. Morgan, *Studies in British Financial Policy 1914–25* (1952), pp. 64, 77, 264–6; A. Vallance, *Very Private Enterprise* (1955); 'Amalgamations and new issues', *Economist* (25 Dec. 1926), p. 1120; Grant, *Postwar Capital Market*, pp. 143–5, 155.
[21] *Economist* (6 Dec. 1919), pp. 1029–30.
[22] P. W. S. Andrews and E. Brunner, *Capital Development in Steel* (Oxford, 1951), pp. 159–61. H. Levy, *The New Industrial System* (1936), p. 203. Morgan and Thomas, *The Stock Exchange*, pp. 206–7.
[23] Collin Brooks (ed.), *The Royal Mail Case* (1933), pp. xvii–xviii.

While there were undoubtedly some legitimate and financially sound merger issues in booms, the new investors in many of them were inevitably disappointed. Postmortems usually revealed the wisdom of the original owners in selling out at boom values and in many cases laid bare the absence of any managerial rationale to the mergers that had been promoted. A number of the postwar boom promotions in the iron, shipbuilding, jute, glass and cotton industries (promoted in the main by Sir Edward Edgar's Sperling Combine and Clarence Hatry's Amalgamated Industrials) required wholesale reconstruction, and in some cases they disintegrated completely.[24] The new issue boom of 1928 also produced some spectacular casualties, and the average depreciation on the issues of 1928 was as high as 41–42 per cent by 1931.[25] To the extent that the former family owners had been paid in these shares (or invested the purchase price in similarly depreciated shares) they suffered along with other shareholders, but with their more direct experience of industrial trends, many of them had wisely demanded cash payments and made substantial financial gains from selling out. Even where they merely exchanged a portfolio of depreciated shares for their own companies, they still had the advantage of a more diversified, and hence less risky, investment. Of course all businesses carried with them risk in varying degrees, but it was extremely unlikely that a diversified portfolio of quoted company shares would do anything other than reduce this risk.[26]

The upshot of the new issue and merger booms was a substantial increase both in the total number of companies quoted on the stock exchange and in their total values. While there were only 569 firms in domestic manufacturing and distribution quoted on the London stock exchange in 1907, the number rose to 719 in 1924 and by 1939 had reached 1712.[27] Over the same period the market values of these quoted

[24] Lord Aberconway, *The Basic Industries of Great Britain* (1927), pp. 202, 231. *Economist* (28 Sept. 1929), pp. 576–7. J. R. Parkinson, *The Economics of Shipbuilding* (1960), pp. 34–5. *Sperling's Journal* (1919–21).

[25] 'The results of the 1928 new issue boom', *Economic Journal*, vol. 41 (1931). R. E. Harris, 'A re-analysis of the 1928 new issue boom', *Economic Journal*, vol. 43 (1933). *Economist* (15 Feb. 1930), pp. 363–4. *Economist* (26 July 1930), p. 182. This performance was worse than the average for quoted companies during the world slump.

[26] If management divorced from ownership was less efficient, or pursued goals different from those of the shareholders, the wealth holders' returns on their investments might, however, be prejudiced.

[27] P. E. Hart and S. J. Prais, 'The analysis of business concentration: a statistical approach', *Journal of the Royal Statistical Society*, series A, vol. 119 (1956), p. 154.

securities rose more than fivefold, from under £500 million in 1907 to more than £2500 million in 1939.[28] Already by the early 1920s, when 57 per cent of profits originated in public companies, those quoted on the London stock exchange included the majority of the more important manufacturing firms, and by 1951 a Board of Trade inquiry reported that quoted companies accounted for some 71 per cent of profits then generated by the corporate sector.[29] The great increase in both the range and the representativeness of stock exchange securities, which had enabled the former owners of family firms to diversify their holdings, also facilitated the formation of greatly enlarged manufacturing enterprises. In the interwar years, for example, we can safely say that enterprises which were valued on the stock market at £32 million or more would effectively have been beyond the resources of all but a few individual capitalists. Without the facilities of the stock market for aggregating wealth, then, such companies could not have been formed. In 1919 there was only one such 'giant' firm, the J. & P. Coats sewing cotton combine, in which the family owners had been pioneers in diluting their ownership by a flotation of their capital three decades previously. By 1930, however, a further six companies – Unilever, Imperial Tobacco, Imperial Chemical Industries, Distillers, Courtaulds and Guinness – had attained this 'giant' size and the number of such companies continued to increase thereafter.[30]

The growth of these giant companies – and also of others in lower ranking (but still large) size categories – was in part the result simply of joining together existing assets which had previously been controlled by separate firms. However, some of their growth also derived from new capital issues to finance real growth in assets, a facility which the stock market could offer on favourable terms to established companies of this kind. Such capital issues were particularly useful to firms which did not have the cash flow to finance large and lumpy investments (such as frequently arose in mass production technology), and thus if they had not had this facility they could not easily have gained access to scale economies. Ford, for example, raised more than £4½ million

28 Calculated from the Stock Exchange *Daily Official List*. Prices increased between 1907 and 1939, but, since they less than doubled, the rise is not mainly due to price inflation.

29 Colwyn Committee, *Minutes of Evidence*, qq. 8550–1. National Institute of Economic and Social Research, *Company Income and Finance 1949–53* (privately printed, 1956), pp. 7–8.

30 See pp. 121–2. The change is a real one, and not the effect of inflation, for there was no firm in 1919 which approached the size of J. & P. Coats and share prices rose by only 15 per cent between 1919 and 1930.

in 1931 to finance the construction of the Dagenham motor car factory complex.[31] Like private firms, however, the large quoted firms of the interwar period relied principally on internally generated funds to finance their new capital projects. Between 1920 and 1938 some 28 per cent of earnings were on average ploughed back into new investment by British manufacturing firms,[32] and in 1924 it was established that some four-fifths of investment in home industry and commerce was financed by such business savings.[33] The system of self-financing had many advantages, particularly in avoiding the transactions costs and uncertainty of new capital issues in the stock market. Hence this source of finance was widely approved by industrialists,[34] and it was also tolerated by shareholders, presumably because of the tax and other advantages and because of the absence of a takeover mechanism to enforce higher payout ratios (see pp. 149–51).

Now self-financing can be an efficient system of allocating financial titles to resources in an economy where there is a direct relationship between the past and future qualities of management and between past profits and future investment opportunities. In this case firms which have in the past performed well will, from their own profits, gain a large portion of total investment resources, and this may reflect the best use of those resources providing the highest social return. However, in a growing economy, with inter-industry shifts and single-industry firms, self-financing is likely to create immobilities, inhibiting the movement of capital to those new investment projects in which the marginal rates of return are highest. This had not been too serious in an economy of small firms where marginal adjustments in the capital stock by many firms could achieve substantial aggregate shifts. However, if only a few large firms dominate a number of industries in which substantial contractions are required, and there are only a few small firms in industries in which large new investments are required, the problem of immobilities

[31] G. Maxcy and A. Silberston, *The Motor Industry* (1959), p. 162. See also E. Nevin, *The Mechanism of Cheap Money* (Cardiff, 1955), pp. 246–8.

[32] P. E. Hart, *Studies in Profit, Business Saving and Investment in the United Kingdom, 1920–1962*, vol. I (1965), pp. 119–20.

[33] H. Clay, 'The financing of industrial enterprise', *Transactions of the Manchester Statistical Society* (1931–2), pp. 213–15. The estimate, which is based on W. H. Coates's evidence to the Colwyn Committee, is only a very approximate indicator. H. W. Richardson's stronger assertion (*Economic Recovery in Britain 1932–39* (1967), p. 149, but cf. p. 201), that in the interwar years new issues were a more important source of funds than either before or since, is problematical, for the historical data on the flow of funds are obscure.

[34] *Macmillan Evidence*, qq. 1537, 8746–52. *Investors Chronicle* (24 Apr. 1937), p. 1184.

could become a serious one. Companies recognizing this could and did overcome the problem by portfolio investment through the market in the quoted securities of other firms when the expected returns were high.[35] Another possibility was to diversify the firm's interests to include related fields and thus participate in the management as well as the finance of other firms. In the car components industry, for example, Joseph Lucas, which had surplus funds at its disposal, expanded its initially limited financial commitment to A. Rist Ltd into a controlling interest and ultimately in 1934 to full ownership.[36] The costs of such relationships were probably lower than for general investments through the stock market by the firm's shareholders, for several reasons in addition to the transactions costs of new issues (which could be high). Firms in the same or in a vertically related industry possess knowledge of each other's manufacturing skills, credit status and market position as part of their working stock of commercial intelligence, so that the costs to them in acquiring information on which to base decisions to acquire share participations are likely to be smaller than the costs to the investor. Furthermore the stock market was clearly ill equipped to judge untried and especially science based investment projects in which future profit levels were the relevant factor in the assessment and past profitability could provide no guidance.[37] The commercial and technical intelligence departments of companies like ICI, by contrast, could perform such assessments more expertly, since they had closer knowledge than the institutions of the stock market of both the relevant production technology and the potential markets.[38] Large firms such as Vickers, GEC and ICI thus received a succession of proposals in technologies related to their own major fields of interest, from individual inventors and from smaller companies which needed the capital resources of a larger company for further development and expansion. Hence the continued flow of resources for new development was

35 Grant, *Postwar Capital Market*, pp. 196–7. W. J. Reader, *Imperial Chemical Industries: A History*, vol. 1 (1970), pp. 384, 421. 'Rationalizing the investment portfolio', *Economist* (5 Apr. 1930), pp. 779–80.

36 Monopolies Commission, *Report on the Supply of Electrical Equipment for Mechanically Propelled Land Vehicles* (1963), p. 26.

37 Committee on Scientific Research of the Economic Advisory Council, *First Report* (1937) (typescript in the *G. D. H. Cole Papers*, Nuffield College Library, Oxford). Grant, *Postwar Capital Market*, p. 279.

38 These generalizations are based on the author's examination of the files of the commercial, technical and development departments of a number of large companies. A fuller and quantitative study of the relative efficiency of large firms, the stock market and individual investors in financing the innovative process would be required to establish the point convincingly.

assured not principally by the stock market (which was generally unwilling to finance untried enterprises) but either by these large firms, or by the time honoured methods of nineteenth century innovators, using private sources of finance.[39] In the case of large firms such projects often laid the basis for future corporate growth and diversification, and in this sphere, as in others, such financial factors were adding to the pressures towards larger firms.

For some smaller firms wishing to achieve the advantages of rationalization and large-scale production, however, neither large companies nor the stock market could supply the required financial resources. As the experienced accountant, Sir Mark Webster Jenkinson, explained to the Macmillan Committee, 'Some have got financial millstones round their necks, and do not know how to bring about fusion. Nobody will take them in.'[40] Often in such cases it was the banks and other creditors who provided the financial pressure, and the financial resources, for rationalization, sometimes against the will of the owners. Though the major banks frequently disclaimed responsibility for forcing bank debtors to merge, and officially ruled out joint action as a breach of the confidential customer–client relationship,[41] they did in fact participate in some enforced reconstruction schemes. In the steel industry, for example, Barclays forced the unwilling directors of Bolckow Vaughan to accept a bid from Dorman Long by the simple expedient of making the renewal of the company's £1 million overdraft conditional on the consummation of the merger.[42]

Such action by the banks was given a more formal structure with the creation, on Bank of England initiative, of the Bankers' Industrial Development Company (BIDC) in 1929.[43] This was intended to devise schemes to re-equip, and where necessary to amalgamate, companies in the staple industries which were often in financial difficulties. It appears to have been primarily an attempt to head off possible government

[39] D. Finnie, *Finding Capital for Business* (1931). 'How new industries grow', *Planning*, No. 68 (Feb. 1936).

[40] *Macmillan Evidence*, q. 3700; see also qq. 7976–9. Andrews and Brunner, *Capital Development in Steel*, pp. 349–60.

[41] *Macmillan Evidence*, qq. 1869–74, 1950, 1977, 2203, though cf. qq. 2388–9. For a complaint by Steel-Maitland of the timidity of bankers and particularly of McKenna of the Midland, see *Baldwin Papers*, vol. 29, pp. 54–63.

[42] J. C. Carr and W. Taplin, *A History of the British Steel Industry* (Oxford, 1962), p. 449.

[43] Sir Henry Clay, *Lord Norman* (1957), ch. 8. A. F. Lucas, 'The Bankers' Industrial Development Company', *Harvard Business Review* (1930), pp. 270–279.

intervention in the financing and reorganization of industry,[44] and, with financial backing from the Bank of England and other City institutions, and a high powered staff, it aimed discreetly to catalyse the banks into action which was felt to be in their long-term interests.[45] An essential part of the BIDC's philosophy was that its capital should in no cases be used to relieve investors of existing financial burdens. Existing assets could not be bought for cash, and thus mergers involving cash acquisitions could not occur under its auspices.[46] Instead shareholders and creditors (including the banks) had to agree to pool their interests in return for a paper title to a share in any future benefits of the merger. Any new capital introduced into such schemes by the BIDC – which, of course, had prior rights – was intended to finance new investment only, and it was on these conditions that the bankers cooperated, some of them somewhat reluctantly. The most spectacular example of the BIDC's work was the formation of the Lancashire Cotton Corporation, which between 1929 and 1932 absorbed almost a hundred firms.[47] The majority of these were forced by their bankers, on the threat of withdrawal of overdraft and loan facilities, to accept the terms offered by the Corporation, though the bankers in fact delayed the process, ever hopeful that by dilatory quibbling they would gain better terms.[48] Though it is conceivable that mergers would have occurred in the industry in the absence of the BIDC (as they did in the fine spinning section of the industry),[49] the Corporation was undoubtedly larger and the operation more swiftly executed than would have been the case if the matter had been left to the directors of the individual companies. Smaller scale mergers were also encouraged by the BIDC in the steel industry, though they were, as the historians of

---

[44] Clay, *Lord Norman*, p. 358. PRO/BT/56/14.

[45] It was seen in the City as an essentially temporary expedient and it was hoped that the financial burden which the banks had shouldered would eventually be floated off to the public when conditions were favourable; see *Macmillan Evidence*, qq. 828–59, 9146. The company was known in the City as B.I.D.: 'Brought in Dead'.

[46] See the debate on the 'Hammersley Scheme' for the cotton industry, which violated this principle and therefore failed to gain support, in *Lloyds Bank Monthly Review* (Oct. 1930, Feb. and Mar. 1931).

[47] Lancashire Cotton Corporation Archives, Annual Reports and Reports of Extraordinary General Meetings, 1929–33. The more generally quoted figure of 140 is incorrect: it refers to projected acquisitions and to *mills* not firms.

[48] *Macmillan Evidence*, qq. 1511–25. *The Times* (30 Feb. 1930), p. 13. Lancashire Cotton Corporation Board Minutes 1929–30.

[49] R. Robson, *The Cotton Industry in Britain* (1957), pp. 158–9.

the industry comment, in large part due to the initiative of individual companies and only partly to Montagu Norman's promptings.[50]

These developments in the stock markets and in banking institutions, then, both liberated firms from the size constraint previously imposed by the wealth of the individual owner and also created new pressures towards larger scale enterprise. More generally, economies of scale in finance were added to the other economies of large-scale operations, and these may have been among the more important scale economies available to firms in some industries.[51] One of the advantages of scale came in the spreading of risks. This may seem to be a superfluous advantage, since we have already seen that capitalists were spreading their risks by diversifying their shareholdings over a wider range of companies. For directors and managers, however, risks were usually concentrated in the companies for which they worked, and since they, rather than the shareholders, took the major entrepreneuria ldecisions, the spreading of risk by diversifying their companies' interests remained an important consideration. More straightforward scale economies were created by the high transactions cost of raising capital by new issues on the stock market. The difficulty of raising capital in amounts of under £200,000 – the 'Macmillan Gap'[52] – was well known and must have inhibited many small firms from seeking a public issue. The basic expenses of issue – prospectuses, advertising, and the professional fees of accountants, brokers, bankers and solicitors – amounted to five figure sums and, as they were fixed costs, this could amount to as much as 20 per cent of the sum raised in a small issue.[53] Furthermore, the securities of small companies were less well known, less marketable and more risky investments, and were capitalized accordingly.[54]

Considerations of financial economies of scale appear to have loomed large for those entrepreneurs who perceived that these debt and equity financing advantages could be added to the less spectacular financial economies available to large firms (such as the aggregation of inventories

[50] Carr and Taplin, *History of the British Steel Industry*, pp. 441, 444–7, 536.
[51] There is a growing body of evidence from modern studies of mergers that financial economies of scale have received less attention in the literature than they merit; see, e.g., J. Kitching, 'Why do mergers miscarry?', *Harvard Business Review*, vol. 45 (1967).
[52] (Macmillan) Committee on Finance and Industry, *Report* (Cmd. 3897, 1931), pp. 173–4.
[53] Lavington, *English Capital Market*, pp. 168–9, 223. T. Balogh, *Studies in Financial Organisation* (1947), pp. 294–7.
[54] Henderson, *New Issue Market*, pp. 106–13. *Macmillan Evidence*, qq. 1526–1527, 3950.

and liquidity and the benefits of cheaper overdrafts and internal banking).[55] The existence of such economies could itself induce firms to merge to achieve the appropriate scale. As Sir Josiah Stamp wrote:

> The average small unit will find it difficult to get finance for rationalization either cheaply or at all. But in as much as the agglomeration of such units may be big enough to command public company finance in London . . . the only hope for rationalization in the small unit industries is a further merging on a considerable scale.[56]

The only practical alternative to a public issue for the many small companies which required long-term finance beyond what they could provide from their own resources was to seek acquisition by an established quoted company (which could usually raise the finance more cheaply from its own funds or by a larger stock market issue). Either way the results of the financial economies were to accentuate the tendency towards the greater concentration of assets in fewer and larger firms.

In a developed stock market, it was also easier to reorganize capital into larger units in order to achieve other kinds of scale economies. Because the ownership of companies was increasingly divorced from control, and titles to assets were readily salable on the stock exchange, a more fluid market in company control could develop. Instead of the delicate negotiations between family firms for consolidating their interests, more aggressive techniques of acquiring firms for restructuring an industry were opened up. E. R. Lewis, for example, a stockbroker who had floated the Decca company, suggested to its directors that they buy another record company, Duophone. When they refused he bought Duophone himself, and then, in 1929, suggested that they sell him the Decca company. Though not personally wishing to sell, the Decca directors realized the financial attractiveness of the offer and agreed to pass it on to their shareholders. It was accepted by over 95 per cent of them.[57] Similarly, Lord Leverhulme's offer of £13 10s. 0d. per share for the deferred shares of Knights in 1920 was so attractive that the company's directors, whilst themselves seeing no logic in the merger, advised their shareholders to accept on purely financial grounds.[58]

[55] J. Stamp, *The Relation of Finance to Rationalization* (1926), p. 5. Grant, *Postwar Capital Market*, pp. 186–9. Balogh, *Financial Organisation*, p. 79. A. E. Musson, *Enterprise in Soap and Chemicals: Joseph Crosfield and Sons Limited 1815–1965* (Manchester, 1965), p. 297.

[56] Stamp, *The Relation of Finance to Rationalization*, p. 25.

[57] E. R. Lewis, *No C in C* (1956), pp. 15–19.

[58] C. Wilson, *The History of Unilever*, vol. 1 (1954), p. 247. *Investors Chronicle* (27 March 1920), pp. 196, 306.

Of course, financially profitable takeovers of this kind would often have been welcomed by closely held family companies also, but in the case of quoted companies neither the personal taste for continuing family ownership nor the practical difficulties of transferring titles intervened to complicate the amalgamation of assets into larger groupings.

In the case of Decca and Knights, although control and ownership were divorced, the directors took into account the financial interests of the shareholders who owned the company, and acted accordingly. This was not, however, invariably the case, as is witnessed by the insistence of some company directors on large side payments to themselves as a condition of agreeing to mergers.[59] Such practices were widespread and, had they developed further and affected the everyday running of quoted companies, they would clearly have created serious problems for the owners of industrial wealth. In general, however, a coalescence of the interests of shareholders and directors was achieved by a variety of expedients, drawn from the earlier experience of those family companies which had employed professional non-family managers in their businesses. The salaries of directors and managers were sometimes directly related to profits, and even where this was not formally so it was a common, and well understood, practice for bonuses to be granted to senior men when the company had had a particularly profitable year, thus creating a similar incentive. Nevertheless in some areas the interests of managers and shareholders did not necessarily coincide. The new generation of professional managers and quoted company directors was as likely to come from the universities, the civil service or the professional classes as to have inherited its managerial positions as a family right, and, though few of its members were of humble origins, they were in general less wealthy than the owning families whom they were gradually replacing on quoted company boards.[60] They relied more on their current earnings than on *rentier* interest on inherited capital. Once established in a senior position, moreover, they were jealous of their rank and tended to regard it as a sinecure in much the same way as the former family owners had done.[61] The growth of their firms (and at the same time of their salaries and their prestige) appears to have been a major objective of such managers. The entrepreneurs who erected statues of themselves next to those of

[59] L. Hannah, 'Takeover bids in Britain before 1950', *Business History*, vol. 16 (1974), p. 72.
[60] C. Erickson, *British Industrialists. Steel and Hosiery 1850–1950* (Cambridge, 1959), pp. 48, 188–203.
[61] Liberal Industrial Inquiry, *Britain's Industrial Future*, p. 90.

Nobel and Lavoisier on the new ICI Millbank headquarters overlooking
the Thames were, we may suppose, as much interested in empire
building and the glory it reflected on themselves as they were concerned
about the income of their shareholders.[62] Yet another pressure towards
larger scale enterprise, in this case from the managers themselves, was
thus implicit in the stock market developments of the period.

Whether from the point of view of financial economies of scale or
from the expansionist ambitions of managers, then, the rapid rise of
the corporate economy becomes more intelligible in the light of the
changing structure of capitalist ownership and the emergence of a
professional managerial class. However, these new developments raised
serious misgivings in the minds of many contemporaries. The conclusions of the Liberal Industrial Inquiry (a body which was strongly
influenced by Keynes) typified these doubts about the upper echelons
of industrial management:

> the vast majority of appointments to Boards of Directors are made
> in effect . . . by co-option by the existing directors. Since the duties
> are indefinite and the privileges agreeable, the way is open to
> various kinds of jobbery. . . . A directorship . . . is apt to be awarded
> to influential people who . . . are without technical qualifications
> for the management of the business. . . . We do not think that the
> Boards, as at present constituted, of Public Companies of diffused
> ownership are one of the strong points of private enterprise. There
> is here an important actual and potential element of inefficiency.[63]

Moreover, shareholders and managers did not only have to face these
consequences of the divorce of ownership and control, but, perhaps
more crucially, had to meet the entirely new managerial and organizational problems presented by manufacturing enterprises which were
larger and more complex than any that had previously been experienced.

[62] Of course, high profits may be necessary both to enable the firm to grow and
to reward shareholders, so the objectives of shareholders and managers may
in practice coalesce.

[63] Liberal Industrial Inquiry, *Britain's Industrial Future*, pp. 90–1.

6

Management and the limits
to growth[1]

Be not frighted, trade could not be managed by those
who manage it, if it had much difficulty.
DR JOHNSON, *Letters*, ed. G. B. Hall,
vol. 2 (Oxford, 1892), p. 126.

જીજી

Even amongst the industrialists who supported the principle of rationalization and were convinced that through it firms could gain access to
important scale economies, there were many who were prey to doubts
about the personal capacities of the men available to run large-scale
enterprises. 'The most difficult thing at present', the Macmillan
Committee was told by a banker, 'is to find a man who can control
10,000,000 spindles. Find that man and I think you will find five or
six positions clamouring for him.'[2] As Sir Alfred Mond told the House
of Commons in 1926:

> The essential of the matter is 'management'. I have come to the
> conclusion that it is impossible for any human being efficiently
> to control any industry beyond a certain magnitude. At a certain
> point they begin to show the paralysis of red tape; they become
> so big that they are like a government department. In my view it is
> impossible to organize industries on a national basis and keep them
> efficient.[3]

This stress on the dangers of bureaucracy implied a belief in individualism in management, an idea that remained an important part of

[1] This chapter is an abbreviated version of an earlier article by the author on
'Managerial innovation and the rise of the large-scale company in interwar
Britain', *Economic History Review*, vol. 27 (1974).
[2] *Macmillan Evidence*, q. 2356.
[3] Quoted in *Business* (Aug. 1931), p. 55. See also E. A. G. Robinson, 'The
problem of management and the size of firms', *Economic Journal*, vol. 44
(1934).

the self-image of the business élite. If Mond himself was able some six months later to embark upon the creation of Imperial Chemical Industries, the largest merger in manufacturing (by market valuation) between the wars, it is nonetheless the case that the fear of diseconomies of scale deterred lesser men.

That a merger did not automatically push back the barriers of managerial diseconomies of scale was a problem often skipped by those advocating rationalization, but central to the concern of contemporary businessmen and observers. Though there are some increasing returns to relative size (in that increased market control both raises profits and reduces uncertainty, freeing entrepreneurial time for other tasks),[4] past and contemporary experience suggested that considerable managerial difficulties would be encountered when companies were merged. As the *Economist* sceptically remarked:

> The advocates of concentration and combination ... are accustomed to dwell on the advantages in respect of efficiency and economy of production and distribution which are derivable from the promotion of standardization of output and specialization of works, the establishment of uniform costings systems, the interchange of information, the combined research, the collective buying of raw materials, and the joint marketing which are thereby facilitated. That such advantages are so derivable is beyond question; but, despite some very striking instances, the fact that they have been generally so derived is still far from established.[5]

Management was the crucial factor in the realization of economies of the type relating to the relative efficiency of firm and market in integrating economic activities. The transition from market relations to intra-firm organization did not occur costlessly and automatically with an increase in scale: it required considerable investment of time, capital and skill in the creation of an efficient administrative structure. Only firms with this organizational investment capacity could embark on an extensive and sustained programme of expansion with reasonable prospect of success.

The reality of this awareness among businessmen and its effect as a

---

4 Cf. the remark attributed to Henry Ford in *Business* (Sept. 1928), p. 129, that 'time given to the study of competition is time lost for one's own business'.

5 'The trust movement in Great Britain', *Economist* (9 Feb. 1924), pp. 240–241.

constraint on the growth of the firm is shown strikingly in the business records of large corporations:

> I contend [wrote a senior manager of Courtaulds on a proposal to acquire other rayon companies] that our present Board and *our present organization* is and will be for a considerable time incapable of running a monopolized industry.... Before therefore proceeding with the idea of trying to obtain control of either the viscose or acetate producers or both, I consider that we should put our own house in order.[6]

On this advice the programme of acquisition was abandoned and it was not until 1957 that Courtaulds acquired its main competitor, British Celanese. Again, when the government was investigating the possibility of mergers in the light castings trade, it naturally looked to the largest merged firm in the industry, Allied Ironfounders, to act as a managerial nucleus, but, here too, the original merger of 1929 had imposed so great a managerial strain on the company as to rule out further acquisitions for some years. 'Any further concentration', it was reported, 'could probably best be carried out through this Company but they feel that it would be unwise to start negotiations for this purpose until they have consolidated their present position.'[7]

Two pieces of evidence, however, counsel caution in the use of managerial diseconomies of scale as a blanket explanation of the cautious approach of some entrepreneurs to mergers and company growth in the interwar years. First, there were a number of very large companies in various industrial fields which showed no evidence of chronically overstretched management. It is clear that organizations such as the Post Office (with 200,000 workers) and the LMS Railway (with 250,000 workers) could function efficiently at a large scale, and, as W. H. Coates of ICI (which employed about 50,000 people in the UK) pointed out:

> There is no industrial combine, so far as I am aware, registered under the Companies Acts which has a number anywhere approaching those figures. The speeches of the Chairman of the LMS, and also the results of the Company notwithstanding statutory and other conditions under which they have to work, show that you

[6] Quoted in D. C. Coleman, *Courtaulds: An Economic and Social History*, vol. 2 (Oxford, 1969), p. 238.
[7] PRO/BT/56/37, Office of the Chief Industrial Adviser, 'Industrial reorganisation', p. 27.

can have efficient management for such an organization. It is hard
to say that those economic units are too large.[8]

Second, and clearly related to the first point, there was a trend towards
larger firms in many industries between the wars, and many of them
could claim, with the Distillers Company, that 'this company has been
a series of amalgamations. Its birth was the result of an amalgamation
and the company has gone on amalgamating ever since.'[9] In the light
of such evidence it would appear to be more realistic to postulate a
managerial limit to the *rate of growth* (rather than the size) of the firm,
a modification which is gaining acceptance in the theoretical literature.[10]
Where managerial skills were highly developed it was realized that the
managerial constraint on growth need not be a significant one at all.[11]
The important variable was management. The analysis of the means by
which barriers to growth were pushed back, as the skills of companies
in digesting acquisitions and in managing large extended organizations
were evolved, therefore offers an important key to the merger process
and the internal development of the modern firm.

The earlier experience of rapidly growing firms had not been promising. A very large proportion of the multi-firm consolidations of the
turn of the century had encountered severe managerial problems,
and this can hardly be a matter for surprise. In the majority of them there
was no obvious 'parent' firm to act as a nucleus or 'core' for the new
managerial structure, for all of the firms involved were, prior to the
merger, of only small size. The rate of growth implied for the largest
firm in the consolidation (i.e. the relation between its size and the sum
of the sizes of the other firms) was, then, perhaps twenty times or more;
that is, of an order of magnitude quite different from that experienced
by firms in normal circumstances. The managerial stresses incurred by
firms undergoing these rates of expansion were documented fully by
the contemporary Fabian civil servant Henry Macrosty and have recently been re-examined by Professor Payne.[12] They need not, then, be
laboured here. Almost invariably their promoters failed to pay sufficient

[8] Management Research Group Minutes (27 Feb. 1935), pp. 12–13 (*Ward Papers*).
[9] Quoted in R. Wilson, *Scotch, The Formative Years* (1970), p. 396.
[10] E. T. Penrose, *The Theory of the Growth of the Firm* (Oxford, 1959). G. B.
Richardson, 'The limits to a firm's rate of growth', *Oxford Economic Papers*,
vol. 16 (1964).
[11] P. S. Florence, 'Reply', *Economic Journal*, vol. 44 (1934).
[12] H. W. Macrosty, *The Trust Movement in British Industry* (1907). P. L. Payne,
'The emergence of the large-scale company in Great Britain, 1870–1914',
Economic History Review, vol. 20 (1967).

attention to the problems of organizing large-scale enterprise and of integrating formerly independent and competing units. As a result many of the new combines were run more as debating societies than as industrial firms: one firm, the Calico Printers' Association, for example, had eighty-four directors, eight of them 'managing', and inevitably suffered from conflicting leadership. Such boards were too unwieldy to be effective, and the introduction of standard costings systems and other methods of maintaining efficiency was too often ignored.[13]

By the interwar years, then, the somewhat light-hearted abandon with which managers had entered upon these early merger agreements had been tempered by a more widespread awareness of the difficulties created by multi-firm consolidations and rapid growth, and large multi-firm mergers ceased to play a significant role in increasing concentration. This can be seen in Table 6.1, which shows the extinction of multi-firm mergers involving twenty or more firms, and an overall reduction in the average size of multi-firm mergers (defined as those involving five or more firms), in the first four decades of this century. The later multi-firm mergers were, moreover, small not only in terms of the numbers of firms involved but also in terms of their capitalization.[14] This abandonment of the large multi-firm merger did not, however, mean that fewer firms were being absorbed by merger in the interwar period; on the contrary the number of firm disappearances by merger actually increased, and in the interwar years overall perhaps as many as 4000 firms were absorbed in mergers and acquisitions (see Appendix I). What had happened was rather a change in pattern: firms, instead of seeking to convert an industry instantaneously into a monopoly, were choosing instead the path of sequential acquisition of smaller competitors and selective mergers with large ones, spacing out their growth more evenly over the years. This more balanced pace of growth not only enabled firms to bypass the chronic managerial stresses encountered by earlier multi-firm consolidations, but also offered the additional attraction of evading the opprobrium which had earlier accrued to the 'Soap Trust' and other attempts to create a monopoly

[13] Only two of the large multi-firm consolidations of the turn of the century – Bradford Dyers, and Fine Spinners – are singled out by Payne ('The emergence of the large-scale company', p. 530) as having efficient management. Others, which involved fewer firms and started with a more promising managerial core firm (e.g. J. & P. Coats, and Imperial Tobacco), were also relatively successful.

[14] Forty-eight out of the seventy multi-firm mergers between 1919 and 1939 involved only five, six or seven firms, and the great majority of these had a combined capital of under £500,000.

instantaneously, rather than by gradual (and less noticeable) piecemeal acquisitions.

The significance of this new strategy in facilitating growth without managerial collapse is underlined by the experience of the odd man out in Table 6.1: the Lancashire Cotton Corporation. This consolidation of 1929 was disliked by many cotton mill managers but was forced on a largely unwilling industry by creditors and bankers acting through the

TABLE 6.1   *The decline of the multi-firm merger in UK manufacturing industry, 1880–1939*

	Mergers involving 20 or more firms	Mergers involving 5 or more firms	Average number of firms per multi-firm merger
1880–9	1	5	19·6
1890–9	6	28	14·6
1900–9	4	20	14·6
1910–19	2	24	10·4
1920–9	1	40	9·1
1930–9	0	18	6·9

Source: L. Hannah, 'Mergers in British manufacturing industry, 1880–1918', *Oxford Economic Papers*, vol. 26 (1974), p. 14.

Bankers' Industrial Development Company.[15] In its first year it acquired seventy companies and in the following year a further twenty-six, making it not only the largest merger (in terms of the number of firms disappearing) between the wars but the largest on record in Britain at any time. The problems of integrating many small and formerly independent cotton mills within one managerial organization were formidable. By mid 1931 it was evident to the financial backers of the Corporation that all was not well and an accountants' investigation into the management and organization was ordered. The resulting report was highly critical of the management, and these criticisms were re-inforced in a further confidential report by Sir Eric Geddes, who had been sent to investigate by the BIDC. Geddes had gained a reputation as a skilled organizer by rebuilding the fortunes of the Dunlop Rubber Company after its collapse in 1921, and he was himself a keen enthusiast for large-scale organization and the rationalization of industry, a passion

[15] See pp. 73–5 above. The account of the Lancashire Cotton Corporation which follows is drawn from an unpublished study of that company by the author, based on the archives of the company and of the BIDC.

he shared with the managing director of Lancashire Cotton, Captain John Ryan. Yet, when he surveyed the company, Geddes could not conceal his belief that the task set for management by the initial amalgamation was superhuman. He expressed these reservations in a letter to BIDC:

I think the founders of the Lancashire Cotton Corporation under-estimated the enormous difficulty of creating a great amalgamation of this kind without any real strong existing organization to take over control.

It would have been difficult enough for any large organization with a loyal and tried staff to take over the industrial properties which have been acquired, but the task was immeasurably greater when one realizes that personnel had to be collected and a central organization created at the same time as the troublesome absorption of the Mills was taking place. It is easy to underestimate this trouble.[16]

Though this was wisdom after the event, there could be no more impressive testimony to the reality of managerial diseconomies of rapid growth than this assessment by a convinced rationalizer and successful entrepreneur when confronted with the results of a multi-firm merger which had ignored the dangers inherent in such consolidations.

For the majority of companies, however, the pace of growth was less hectic than this, and there were remarkably few mergers in the 1920s and 1930s which involved more than a doubling in size for the 'core' firm. Interwar entrepreneurs thus in general avoided the mistakes of their predecessors in this area, but they did not, on that account alone, banish all the managerial problems of rapid growth and large scale. The nature of these problems can be seen from the analyses of the coordination difficulties met by large organizations, which have been made by managerial theorists.[17] It is generally accepted that an executive can conveniently control only a limited number[18] of subordinates: expansion cannot efficiently be achieved by adding to his span of control above this number; instead an extra hierarchical level must be added.

---

[16] Letter, 7 Dec. 1931, from Sir Eric Geddes to E. R. Peacock (Bank of England Archives, Securities Management Trust File).

[17] For a review of this literature, see O. E. Williamson, 'Managerial discretion, organisational form and the multi-division hypothesis', in R. Marris and A. Wood (eds), *The Corporate Economy* (1971), pp. 343–86.

[18] Six is often suggested as the optimum, though variables such as personality, team spirit and the complexity of the decision situation will, of course, permit a higher number (or dictate a lower number) in particular circumstances.

D

However, a very large firm employing this principle needs a long hier-archy of command, and control loss increases as the chain of command is extended, partly because of the decreasing efficiency of communica-tion, and, further, because of the expanded field of discretionary be-haviour for subordinates which is introduced. In some of the lower reaches of the firm, the adoption of standardized practices might help to widen the span of control (and thus shorten the lines of command), but at the level of senior management this is clearly difficult in view of the strategic and tactical decision making required at those levels, especially in a growing firm. Generally, therefore, an increase in size will expand the strain on an organization and limit its efficiency.

Technical progress was reducing some of these problems of communi-cation and control in large corporations between the wars. In particular the telephone, perhaps the most important new instrument of communi-cation to become widely available to managers, provided a means of rapid communication between departments, or between geographically dispersed branches, and thus facilitated managerial control.[19] Technical advances substantially cheapened and improved the quality of telephonic communication, and the number of business telephones doubled between 1914 and 1930 and continued to rise rapidly in the 1930s.[20] In the case of other kinds of office machinery the rate of adoption was perhaps even more dramatic. Before the war such machines had been little used in business. In the works office of Stewarts & Lloyds in 1903, for example, it was later recalled that:

The ¦only piece of modern equipment was the telephone, with a private line to the Glasgow Office. . . . But there was no other office machinery of any kind – no typewriters, adding machines, comp-tometers, pay-roll listing machines, etc. There were no women in the office. Everything was handwritten and the only duplication was by letterpress copying. . . . No one trained in a modern office can have any idea of the crudity of office methods and organization at that time.[21]

In the interwar years, by contrast, typewriters, duplicators and account-ing machines of various kinds were widely used in business, and the

[19] A. Marshall, *Industry and Trade* (4th edition, 1923) pp. 229, 363, n. 2. *Textile Recorder* (15 May 1919), p. 29.
[20] Hannah, 'Managerial innovation and the rise of the large-scale company', p. 257.
[21] Quoted, from *Stewarts and Lloyds Limited 1903–1953* (1953), p. 15, in Payne, 'The emergence of the large-scale company in Great Britain 1870–1914', pp. 534–5.

mechanization of routine information gathering and processing was helping to overcome the problems of information presentation, and hence of coordination and control. While none of the interwar office machinery innovations had the same discontinuous effects on information and management in the large corporation as the computer (which was later developed from them), the claim of Powers-Samas in 1930, that 'without ... mechanization of office accounting the rapid growth of business and the formation of large consolidations would have been difficult if not impossible',[22] clearly contained more than a grain of truth. Most important of these innovations were machines such as the Hollerith which could process accounting data with great speed, and which facilitated significant improvements in the collection and diffusion of information in large companies.[23] Furthermore, the effect of mechanization was not only to cheapen information processing, but, often more significantly, to stimulate new thought about systems of management control. At a Management Research Group meeting in 1937, for example, 'it was generally agreed by members after a short discussion [on office machinery] that of the saving in cost 10 per cent was effected through mechanization and 90 per cent by the general overhaul of existing systems'.[24] It was through such structural changes in management that the more significant organizational changes of these years had their impact on the growth of the firm.

The first reaction to the stresses of growth on company management was often the functional differentiation of managerial tasks. The increasing division of labour in management is reflected in the growth of specialized professional associations and institutions and the expansion in the employment of specialist managers.[25] Most of the large companies which were members of Management Research Group Number 1, which had been founded in 1926, could send functional specialists occupying senior central office positions to group meetings discussing

[22] Advertisement, *Business* (Oct. 1930), p. 152.
[23] L. R. Dicksee, *Office Machinery and Appliances* (1928), p. 177. 'How Holleriths came to Britain', *Tabacus* (1957). 'Machinery in the office', *Economist* (26 Feb. 1938), pp. 431–2. C. H. Costello, 'A quarter century's progress in management method and business equipment', *Business* (Jan. 1933), p. 31–3, 44, 47. A. P. Hodges, 'Development in office machinery and equipment', in R. Pugh (ed.), *British Management Year Book* (1939). A. E. Musson, *Enterprise in Soap and Chemicals: Joseph Crosfield and Sons Limited 1815–1965* (Manchester, 1965), p. 293.
[24] Management Research Group Minutes (10 Nov. 1937), p. 3 (*Ward Papers*).
[25] H. Whitehead, 'The changing environment of management', *British Management Review*, vol. 6 (1947).

specialized topics in personnel, finance, accounting and technical matters.[26] The recommendation to 'spread overheads' in much contemporary management literature refers, among other things, to the managerial economies of scale available in the use of such specialists, and also to the wider use of information, the cost of acquiring which did not increase with firm size.[27] The proliferation of functional specialists was important in that it enabled a central office to delegate managerial functions of an advisory or routine kind to specialists, whilst itself concentrating the efforts of entrepreneurial peak coordinators on the initiation and planning of general business policy and the efficient oversight of the manufacturing divisions of the firm.

Whilst such functional specialists could be trained in the new specialisms on the job, there was clearly a problem in recruiting as peak coordinators entrepreneurs and managers with experience of large-scale bureaucratic administration. Unlike the small firm the large corporation was unable to rely exclusively on traditional sources of new recruits through social and family ties but had to develop a wider net of management recruitment.[28] The problems of managing large enterprises in manufacturing did, of course, have something in common with those of other large organizations, but there appears to have been little managerial spin-off from, say, the railways to the manufacturing sector. Indeed in the twentieth century such spin-off as occurred appears to have been in the reverse direction: Sir Josiah Stamp, for example, who had been a director of Nobel Industries (one of the largest manufacturing companies in the early 1920s), became president of the London Midland and Scottish Railway, taking with him his experience of large-scale administration at Nobel.[29] Originally he had been trained in the Inland Revenue rather than in business, and for some companies the government sector remained the most promising source of recruitment. A steady succession of civil servants from the Revenue followed Stamp

26 Management Research Group Minutes. The companies included ICI, AEC, Standard Telephones & Cables, Metal Box, and Dunlop.
27 J. M. Clark, *Overhead Costs* (1923), pp. 119–23, 141. D. H. Robertson, *The Control of Industry* (1928), pp. 22–6. *Economist* (16 Aug. 1930), pp. 329–30.
28 For criticisms that business did not succeed in attracting the ablest men, see, e.g.: H. J. Habakkuk, *American and British Technology in the Nineteenth Century* (Cambridge, 1962), p. 191; Liberal Industrial Inquiry, *Britain's Industrial Future* (1928), pp. 130–1.
29 J. H. Jones, *Josiah Stamp: Public Servant: The Life of the First Baron Stamp of Shortlands* (1964), pp. 179–86, 281–91. Cf. J. Stamp, 'Administration of business and public affairs', *Journal of the Institute of Public Administration*, vol. 1 (1923).

into the higher echelons of the management of ICI (which had acquired
Nobel in 1926), beginning with W. H. Coates, the right hand man of the
ICI chairman between the wars, and continuing with Sir Paul Chambers
and other directors in more recent times. Other large businesses, notably
Vickers, recruited senior management from retiring military personnel,
a practice which had not only the commercially attractive overtones of
pantouflage but also the advantages of recruiting senior men experienced
in bureaucratic procedures and problems of administrative control in a
large organization.[30] However, not all firms could rely solely on such
external sources of expertise, so managerial skills increasingly were
developed in the training of functional specialists and departmental
managers within the manufacturing enterprises themselves. As Stamp
told the Macmillan Committee,

> The kind of experience required in consolidating an industry . . .
> is a thing which, with a certain translation of terms, can be passed
> from industry to industry, and given men of the requisite know-
> ledge and intelligence that experience can be cumulative, that type
> of man can be gradually evolved. . . . There are a few people who
> are born to do it, and who see their way through; there are others
> who could quickly learn from experience.[31]

One avenue of experience commonly followed was an accountancy
training, for it was particularly through developments in accounting
that the introduction of new methods for the oversight and assessment
of subsidiaries was encouraged and facilitated. Cost accounting had a
long pedigree as a means of achieving efficiency through managerial
control in business,[32] but it could still be claimed in the war that British
businessmen had not adequately come to terms with it,[33] and some of
the most useful techniques of industrial accounting, as Josiah Stamp
pointed out in 1929, were 'almost entirely a development of the last
twenty years'.[34] Hence the demand for more rigorous methods of

---

[30] Cf. F. Lee, *Manufacturer* (1938), p. 102. General Sir Ian Hamilton, *The
Soul and Body of an Army* (1921), pp. 230, 239.

[31] *Macmillan Evidence*, q. 4128.

[32] N. McKendrick, 'Josiah Wedgwood and cost accounting in the industrial
revolution', *Economic History Review*, vol. 23 (1970).

[33] L. R. Dicksee, *Business Methods and the War* (1915), pp. 42–3.

[34] J. Stamp, *Some Economic Factors in Modern Life* (1929), p. 159. See, on
accounting generally: N. A. H. Stacey, *English Accountancy, 1800–1954:
A Study in Economic and Social History* (1954); R. H. Parker, *Management
Accounting, An Historical Perspective* (1969); F. R. M. de Paula, *Developments
in Accountancy* (1948).

financial assessment and management information from large businesses between the wars was mediated by a rapid rise in the proportion of accountants working in industry: in 1913 a substantial majority of English accountants were in private practice, but by 1939 over half of them were in the direct employ of business.[35] Accountants, who had previously only rarely taken important positions in the directorates of large companies,[36] now began to play a more significant part. At the Dunlop Rubber Company, for example, F.R.M. de Paula (who had been Professor of Accounting at the London School of Economics in the 1920s) was invited to become Controller of Finance in 1929. He developed a comprehensive system of internal audit, costing and forecasting, and Dunlop's finance division was able to provide information for the control of costs and investment, within a framework of annual budgets for subsidiaries. De Paula himself repeatedly stressed the relevance of these methods to overcoming the managerial diseconomies of scale sometimes encountered by merged enterprises, by securing centralization of control with decentralization of responsibilities.[37] The career of Francis D'Arcy Cooper at Unilever also indicates the expanding role of the accountant in solving the organizational problems of large enterprise. Coming to Lever Brothers from the accountancy partnership Cooper Brothers in 1923, he was largely responsible for consolidating the extensive and disparate parts of the empire built up by William Lever, and, after succeeding the founder as chairman of the company in 1925, he eventually became the chairman of the merged Unilever company and pursued his policy of consolidation and rationalization of subsidiaries there throughout the 1930s.[38]

The nature of the need for and the disadvantages of the centralized control that was being facilitated by these technical and accounting innovations was succinctly stated in 1930 by Lord Melchett (formerly Sir Alfred Mond and then chairman of ICI) when he said that 'the real problem of rationalization and merging of big enterprises consists in effective central control with sufficient elasticity lower down to allow action to be neither arrested nor delayed'.[39] In many companies this was

[35] Stacey, *English Accountancy*, p. 215. De Paula, *Developments in Accountancy*, p. 207.

[36] Dicksee, *Business Methods and the War*, p. 43.

[37] 'The role of finance and accountancy in the management of large business combines' (1933), reprinted as ch. 13 in de Paula, *Developments in Accountancy*, and Management Research Group Minutes (12 Mar. 1934), 'Budgetary control practice', discussion.

[38] C. Wilson, *The History of Unilever* (1954), vol. 1, pp. 273, 297–310; vol. 2, pp. 309–13.   [39] Quoted in *Economist* (3 May 1930), p. 988.

to involve the splitting of activities on a regional or product group basis, and the creation of independent profit centres ('divisions') responsible for their performance to a peak executive, which controlled finance and capital investment. Since Melchett felt that ICI had attained these objects, that company is an appropriate one in which to study the evolution of a decentralized system of subsidiary management in harness with central office control.

At the time of its formation, in 1926, ICI was the largest merger in British manufacturing industry in terms of its capitalization, an amalgamation of four companies all of which had previously been extensively engaged in mergers. The two larger partners – Nobel and Brunner Mond – dominated the explosives and alkali sections of the chemical industry respectively, and, together with the British Dye-stuffs Corporation and the United Alkali Company, they formed a diversified chemicals and metals group which had a market value at the time of its formation of over £60 million.[40] The company continued to grow both internally and by acquisition, and no insurmountable managerial barriers to expansion seem to have been reached. The managerial structure of ICI should, then, offer some indications of the expedients by which large organizations were able to overcome managerial limits both to the absolute size and to the rate of growth of the firm.[41]

Its managerial structure seems to have owed most to the largest partner in the merger, Nobel Industries Ltd. Brunner Mond, which was in the throes of a managerial reorganization at the time of the merger, envisaged a somewhat looser framework of control, but already in the provisional agreement to merge it was accepted that the new company would adopt a Nobel-type structure. Nobel had its origin in the British end of the Nobel Dynamite Trust which, having been split off from its German counterpart in the First World War, amalgamated with its main British competitors – some thirty or more companies in all – to form Explosives Trades Limited, the company changing its name to Nobel Industries Limited in 1920. Centralization of these separate interests was first required and it was for this purpose that the Nobel structure was originally devised. Within six years of the merger, a central research department had been established at Ardeer, and production had been centralized in the most efficient factories, the remainder being closed

---

[40] *Economist* (30 Oct. 1926), p. 721.
[41] The analysis of the managerial structure at Nobel and ICI which follows is based on various published and unpublished sources, listed in Hannah, 'Managerial innovation and the rise of the large-scale company in interwar Britain', p. 260, n. 4.

down and sold off or transferred to other uses. This seems to have been achieved by the efforts of H. J. Mitchell, aided by John Rogers and Josiah Stamp, in creating a central office structure at Nobel House in London which could handle the managerial problems of consolidation and growth. Routine functional responsibilities such as central purchasing, personnel, publicity, legal, taxation and investment matters were centralized at Nobel House. The legal form of control for the original subsidiaries and for new acquisitions (of which there were many) was for Nobel Industries to become the sole director and for a delegate board to be appointed, responsible to the Nobel directors who served on the main working committees of the Nobel Board – the Management Committee and the Finance Committee. The assessment of the performance of the subsidiaries was facilitated by a unified system of merger accountancy which was developed by Josiah Stamp at a central Secretarial Department. This was fully operational by 1923, showing the financial results of all companies in the group on a common basis. Two other central departments – the Development Department under Todhunter (which investigated proposals for diversification and technical innovation) and the Central Executive and Advisory Department under Mitchell (which acted as the board secretariat and supervised trading agreements and commercial intelligence) – had important entrepreneurial roles. In particular they jointly (and sometimes with the assistance of outside accountants) assessed potential acquisitions in their financial, commercial and technical aspects. The skill of their premerger assessments, together with the central postmerger control, facilitated the assimilation of acquisitions and the release of synergy in the mergers which were undertaken. Growth was thus self-sustaining as the financial and managerial 'surpluses' generated were made available for further expansion.

When this centralized system had been successfully established it appears that some devolution of responsibility onto the manufacturing units was attempted, Mitchell aiming at 'the retention at HQ only of the ultimate control and of certain specialized service departments'.[42] The metal manufacturing subsidiaries were organized from Birmingham rather than from Nobel House and, although financial control and rationalization of capacity were enforced, they were kept on a loose rein. Although the major ingredients of a financially centralized group with decentralized divisional management, which later developed in ICI, were present in this structure, they were there only in germ. It was only

[42] Anon., *ICI Ltd and its Founding Companies*, vol. 1 (1938), p. 240.

with the increase in size after the ICI merger that a full system of divisions emerged. Even then, however, full centralization of ICI itself was to precede the programme of decentralization.[43]

Although Sir Alfred Mond became the first chairman of ICI when it was formed in 1926, men from Nobel occupied strategically important positions in the new company. The considerable managerial 'surplus' of the Nobel group (indicated by the reported drying up of attractive investment possibilities in its Development Department's Reports) was now released upon the task of organizing the larger company. Investment management was soon centralized in the Finance Committee, and an Executive Committee, consisting of directors with functional responsibilities, controlled all new capital expenditure and provided a central forum for policy making. Banking, purchasing, commercial, staff, and statistical control policies were very quickly standardized and centralized. In short, an enlarged central office of service departments and strategic entrepreneurial committees was developed similar to that which had obtained in Nobel. The rationalization of production capacity on the basis of technical and financial reports on subsidiary factories had by the late 1920s been applied to the whole range of ICI manufactures. Expansion continued as the manufacture of new products was initiated and new subsidiaries were acquired (after investigation by the central Development Department) in the metals, fertilizer, dyestuffs and general chemical fields. The system of regional sales offices originally developed by Brunner Mond was taken over by the group and became the sales offices for the majority of ICI products.

This centralization brought lower costs through enhanced buying power and rationalized production, joint distribution and selling, better financial control and improved cash flow, and better use of scarce research, commercial intelligence, and other managerial talent. Nevertheless the structure soon showed signs of the weaknesses of centralized decision making, and between 1928 and 1931 manufacturing was decentralized to eight 'groups', each with a chairman and a delegate board consisting of local executives and liaison officers from head office. The model for the groups was the Birmingham end of the business

---

[43] It appears to have been the view of the Nobel directors that such a cycle of decentralization was necessary in a merger, centralization initially providing unified accounting and personnel policies and being followed by decentralization to encourage local managerial initiative and efficiency; see, e.g., J. Stamp, 'The management and direction of industry', 1930 broadcast reprinted in his *Criticism and Other Essays* (1931).

which had been inherited from Nobel and had subsequently grown by the acquisition of further non-ferrous metal companies. Traditionally independent in Nobel, this group was in 1928 reorganized and each of the four companies in it was also decentralized. This was soon recognized as an appropriate model for the larger chemical subsidiary companies also and the other groups were set up in the following year.

Inter-group policy was coordinated by a Central Administration Committee on which the group chairmen sat together with the senior ICI functional officials. Formally, two subcommittees of the main Board, the Finance Committee and the General Purposes (formerly Executive) Committee, exercised ultimate control. Channels of communication were thus established by the common membership by central and 'group' (i.e. local) executives of central committees and 'group boards'. The main Board's authority was exercised through control of capital expenditure on the basis of past results (assessed by the unified accounting systems and interdivisional market pricing which had been introduced during the centralization phase), and of future prospects (assessed by technical and commercial experts). Finance thus remained centralized and, through performance measurement and annual budgeting, controlling.

By the 1930s, then, the company had clearly developed into a modern decentralized corporation with a functionally specialized head office exercising overall financial control and providing managerial and financial services to the divisions ('groups'). It could be seen as a federation of semi-independent firms with the central office providing a highly efficient capital market, management consultancy, and service agency. The logic of such a structure had not by the close of the inter-war period been fully worked out at ICI. The chairman, by now Baron McGowan of Ardeer, maintained autocratic central control and insisted, for example, on central office control over pricing policy, so that individual profit centres were not in fact wholly autonomous, and the 'groups' themselves could not therefore be held fully responsible for commerical results which emanated from company not 'group' policy. The 'groups' were thus concerned principally with efficient works management and it was in the sphere of production rather than of commercial affairs that decentralization was really effective. Nonetheless, for all its limitations the structure clearly approximated to the paradigm of the multidivisional corporation.[44] Aided by more favourable

[44] Cf. L. F. Haber, *The Chemical Industry 1900–1930* (1970), p. 341.

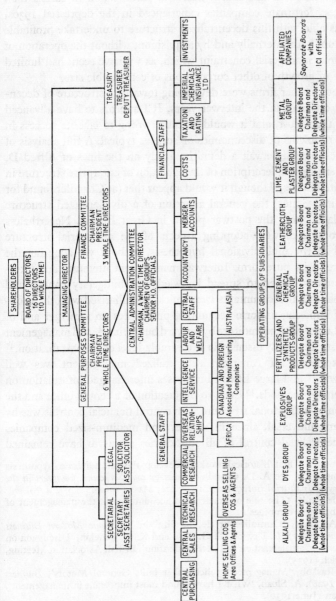

FIGURE 6.1 *Organization chart of Imperial Chemical Industries Ltd, 1935. Source: H. J. Mitchell, 'Methods of inculcating modern management principles and practices in large-scale undertaking', in Sixth International Congress of Scientific Management, Development Section Papers (1935).*

demand conditions and more successful participation in cartels than many less fortunate companies experienced in the depressed 1930s, ICI was able with this decentralized structure to undertake profitable expansion, both internally and by acquisition, without the operation of the serious managerial constraint which, as we have seen, had limited the rate of growth of other corporations of comparable size.

Although other firms were developing towards a structure of decentralized divisions in the interwar years, ICI appears to have advanced further than most, and it would be quite wrong to take its success in solving problems of diseconomies of scale as typical. A full analysis of this question must await a definitive study on the lines of Alfred D. Chandler's classic description of the evolution of enterprise structure in the United States,[45] though it would appear that (as Chandler found for the United States) the general adoption of a divisionalized structure was delayed until the postwar period in Great Britain. Nevertheless some companies were adopting aspects of the managerial structure implied in the multidivisional hypothesis, without yet adopting the structure in its pure form. Interwar management literature and inter-firm contacts helped to spread knowledge of budgetary control and divisionalization. Accountants were quickly familiarized with the problems of imposing uniform accounting procedures on a merger and of controlling capital expenditure by forward budgeting,[46] and management specialists frequently discussed the desirable level of decentralization.[47] The experience of American decentralized management was well publicized,[48] and large British businesses interchanged information on management methods.[49] Even so the creation of a central office and the adoption of a divisional structure with central financial controls was by no means universal, since in smaller and medium-sized companies direct centralized control of all subsidiaries seems to have remained

[45] This was the first historical work to develop the multidivision hypothesis systematically; see A. D. Chandler, *Strategy and Structure, Chapters in the History of Industrial Enterprise* (Cambridge, Mass., 1962).
[46] E.g. de Paula, 'The role of finance and accountancy in the management of large business combines'.
[47] E.g. A. Plant, 'Centralize or decentralize' in his *Some Modern Business Problems* (1937); L. Urwick, T. G. Rose and K. G. Fenelon, 'Discussion on problems of amalgamation and decentralisation', British Association Meeting, Norwich (1935).
[48] J. J. Raskob, 'Management policies that built General Motors', *Business* (Oct. 1928). A. Sloan, 'What I have found most important in management', *Business* (June 1930).
[49] Especially through the Management Research Groups, one of which (No. 1) was reserved for large companies.

possible and decentralization of responsibility was limited.[50] Even among larger companies a decentralized system was adopted as much by default as by choice, as at ICI, after an initial centralizing phase. Where no deliberate strategy of decentralization on the ICI model was developed, the holding company form of control of subsidiaries was most commonly used and internal competition of an administered kind was allowed to persist with only minimal policy and financial controls being exercised from head office. In this sense a decentralized system was the natural form of management in a company growing by acquisition and it was widely adopted,[51] though there was great variation, both between companies and within the same companies at different points in time, in the degree of central control which was exercised. Associated Electrical Industries, which had been formed in 1928 by the consolidation of British Thomson-Houston and Metropolitan-Vickers, continued to run them separately for many years, with loose connections through staff meetings (for exchange of information rather than managerial coordination) and a central financial control which was very much at arm's length.[52] Other companies described as loosely run confederations of subsidiaries with little central control include Tube Investments, Imperial Tobacco, Tootal Broadhurst Lee, Hawker Siddeley, Guest Keen & Nettlefold, and Electrical & Musical Industries.[53] If this looser structure prevented the full realization of all the organizational economies of rationalization, it nevertheless gave them access to the benefits of pooled overheads, risk spreading, the interchange of commercial and industrial methods, collusive pricing policies, and some degree of coordination of new investment. Provided that no diseconomies resulted from their large scale, these firms could grow rapidly by reaping such

[50] 'How far should managers be managed? How we work from head office by the chief executives of 19 firms', *Business* (Feb. 1930), pp. 72–4.
[51] W. G. Hiscock, 'Centralisation or decentralisation', *British Management Review*, vol. 4 (1940), pp. 136–9. Sir Gilbert Garnsey and T. B. Robson, *Holding Companies* (3rd ed. 1936), pp. 18–24.
[52] G. Walker, 'Development and organisation of AEI Ltd.', in R. S. Edwards and H. Townsend, *Business Enterprise* (1958), pp. 309–13. R. Jones and O. Marriott, *Anatomy of a Merger: A History of GEC, AEI and English Electric* (1970), p. 150.
[53] Edwards and Townsend, *Business Enterprise*, pp. 66, 220–7, 293. G. Turner, *Business in Britain* (1969), pp. 52, 352. B. W. E. Alford, *W. D. and H. O. Wills and the Development of the UK Tobacco Industry, 1786–1965* (1973), pp. 309–18, 331–3, 365–70, 445–6. *Economist* (31 Oct. 1931), p. 825. Unilever, especially under William Lever in the early 1920s, was also run as a loose confederation, see Wilson, *History of Unilever*, vol. 1, pp. 267–96.

benefits without encountering any managerial diseconomies of scale or of growth.

However, many firms which failed to develop an appropriate balance of centralization and decentralization failed to create the managerial surplus and profits stream which was necessary if acquisition activity was to be sustained, and the rate of growth of large firms was still significantly constrained by the management variable. Though Nobel could, in a multi-firm merger, double in size in 1918, then continue a programme of consolidating and diversifying acquisitions over the next eight years, and subsequently by merging with three other companies triple in size to become ICI, which then itself expanded by further acquisitions, the majority of companies had to accept a more leisurely pace of growth. Even mergers on a more modest scale than those of the earlier multi-firm mergers could run into grave problems. In 1924, for example, a committee of inquiry into Crosse & Blackwell (a merger of seven companies) had to recommend the writing down of the capital by £2¾ million, finding:

> ... serious duplication and overlapping in management [and] that the benefits which had actually been derived from combination were so few as to be practically non-existent; and, worst of all, that the associated firms had been competing with one another as strenuously as ever.[54]

Those involved in rationalization could still, it seems, give far too little attention to the management factor for the subsequent health of their creations.

On the other hand, the firms which did solve the problems of large-scale organization and of discontinuous digestion of acquisitions were able to achieve high growth rates through merger, involving a doubling or tripling of firm size in a year. Companies like ICI, Reckitt, Fison, Metal Box, British Plasterboard, Distillers, Tube Investments, and many other leading acquirers in the large and medium company size ranges could, and did, grow rapidly between the wars, though they rarely attempted to acquire more than one large or three or four small or medium-sized companies in a year, preferring to keep their rate of growth to a manageable level. Such companies, especially those like ICI which had solved the current organizational problems of merger, were already approaching rationalization, monopolization and company

[54] Quoted in P. Fitzgerald, *Industrial Combination in England* (1927), pp. 193–194.

growth with increasing confidence. They were active in important sectors of the economy, and other industries were exhorted to follow their example. In 1935 a research group of Political and Economic Planning was claiming that:

> Now that the technique of large-scale production is much nearer solution owing to the realization that the advantages of centralized control can only be achieved through functional and administrative decentralization it cannot be overemphasized that the [cotton] industry must face up to the necessity of further amalgamations.[55]

The management developments which we have described were thus both an inspiration of and a response to the merger waves, an integral part of the rationalization movement as well as a condition of its success. As managers solved these problems, economic coordination by competition and the market could begin its retreat. If a misquotation from Burke be permitted, sophisters, economists and calculators were coming in the large decentralized corporation to supplement and partially to succeed the more inchoate signals and disciplines of the market.

By the interwar decades, then, circumstances were in a number of respects more favourable to the development of large-scale business organizations than they had been during the earlier, and sometimes abortive, movements towards higher concentration of the turn of the century. Managers had previously, in smaller firms, often been able to take questions of organization for granted, devoting their attention to marketing and other policy problems. Now they found that they were to a considerable degree involved in organization making. This was a natural requirement of a movement in industry which aimed at supplementing, and to some extent replacing, coordination by the 'Invisible Hand' of the market, by the organization, planning and execution of economic activities within firms. This, in turn, was increasingly desirable as economies of scale and of integration in production technology, marketing and finance were developed, and it was favourably regarded by a benevolent government and by public opinion. The growth of the stock exchange and the transformation in the ownership of enterprise provided the financial wherewithal for the economies to be realized by the grouping of assets into large firms. Thus, whilst evident difficulties remained, increasingly it seemed that the governmental,

[55] Political and Economic Planning, *Report on the British Cotton Industry* (1934), p. 108.

financial and managerial constraints on the growth of firms had been loosened sufficiently to produce substantial growth. The foundations on which the large-scale corporate economy could be built were already firmly laid.

7

The rise of the corporate
economy: dimensions

How much ? how large ? how long ? how often ?
how representative ?
J. H. CLAPHAM, 'Economic history as a discipline',
Encyclopaedia of the Social Sciences,
vol. 5 (New York, 1930).

ಬಬ

The rise of the corporate economy was a gradual, evolutionary process
which had its roots firmly in earlier manifestations of capitalism and
enterprise. It was not – any more than was the industrial 'revolution'
before it – a sudden revolutionary overthrow of established economic
structures and relationships, though it was eventually to transform the
economy from which it grew. This view of industrial evolution is
perhaps uncontroversial, since it accords with the general impression
of the slow growth of enterprises, but it would be wrong to infer from
this the stronger view that the rise of the corporate economy has
progressed evenly and continuously over successive decades. We have
already seen that its birth was troubled and its early growth exceptionally slow. In the interwar rationalization movement, however, there was
a strong and growing body of opinion which believed that its progress
should be accelerated. The aim of the present chapter is to identify
the time pattern of the growth of large corporations which followed,
and in particular to identify the period during which they grew most
rapidly to achieve a state which resembled the corporate economy of
today and diverged significantly from the nineteenth century structure
of enterprise. In order to establish whether such a period of rapid
transformation in the structure of manufacturing industry can be pinpointed, it will be necessary to examine the dimensions of industrial
concentration and of merger activity over the first seventy years of the
present century.

The intrinsic difficulty of measuring concentration for historical
periods in any country is compounded in the British case by the absence
of any definitive statistical work on the subject until relatively recently.

'It may very well be,' Sir John Hicks cautioned his fellow economists in 1935, 'that monopoly is more important today than it was fifty years ago, though it is not so obvious as it appears at first sight.'[1] Even if we confine our attention to the measurement of the sizes and market shares of large firms, rather than the more nebulous concept of monopoly power, the difficulty is no less real. For the period prior to 1935 we lack even the barely adequate statistics of concentration in individual industries which, since the pioneering work at the Board of Trade by H. Leak and A. Maizels, have been regularly published in official sources.[2] In the absence of such statistics, there are dangers in generalizations about aggregate changes based on the limited and biased sources which have hitherto been available. Evidence of monopoly is in some ways more likely to be available on more recent than on historical periods, for the methods by which industrialists have built and maintained monopolistic positions now gain notoriety and are enshrined in the reports of the Monopolies Commission. Yet while the rise of the large corporation is evident, it is often forgotten that concurrently there has been an expansion of markets and a growth of new industries, many of which consist of small and specialized firms. It is well to remember that many important developments in twentieth century industrial practice have counterbalanced the trend to 'rationalization' which we have emphasized. The spread of electric power, for example, has freed industrialists from the scale constraint previously imposed by economies of scale in the generation of steam power, and this has enabled many small firms to survive and grow.

Nevertheless, an overall impression of substantially increasing concentration in the course of the twentieth century is inescapable. Victorian industry in its heyday, whilst it was not without its oligopolists, was typically carried on by small and medium-sized firms. Although there were many large firms, the majority of these were in industries such as steel and brewing where the market was also large, so that most British industries in fact still had an atomistic structure. Even as late as 1907 only a few industries had clearly developed high levels of concentration, and there is general agreement that the overall level of concentration was considerably below that achieved in the United

---

[1] J. R. Hicks, 'Annual survey of economic theory: the theory of monopoly', *Econometrica*, vol. 3 (1935), p. 1.

[2] H. Leak and A. Maizels, 'The structure of British industry', *Journal of the Royal Statistical Society*, series A, vol. 108 (1945). Subsequently to their work, the *Census of Production* statistics have regularly provided data on concentration in individual industries.

States in the period before the First World War. The contrast with the situation two decades later, when more accurate data collected in the 1935 Census of Production throw more light on the structure of industry, is evident. At that date, it is possible to assess the share of the largest three firms in the total output or employment of the majority of industries. This shows that 8 per cent of total employment was then accounted for by industries in which the share of the top three firms in employment was 70 per cent or more.[3] Many of these industries with high levels of concentration – non-ferrous metals, rayon, margarine, dyestuffs, refrigerating machinery, telephone apparatus, petroleum, gramophones, rubber tyres – were 'new' in the sense that they had grown rapidly between 1907 and 1935. There were also many older trades among the concentrated industries: iron and steel pipes and tubes, explosives and fireworks, incandescent mantles, matches, polishes, soap, spirit distilling, sugar and glucose and weighing machines. These also had undoubtedly undergone considerable structural change, often as a result of mergers between old-established firms, in the decades prior to 1935.[4] The evidence thus seems to support the view that by 1935 concentration among firms in United Kingdom industries had risen to a higher level. This view is, moreover, strengthened by comparison with the structure of firms in the United States in the same year. Comparison of concentration ratios on an international basis is notoriously difficult, but the best evidence suggests that by 1935, taking the average over a range of industries, they were at about the same level in both countries.[5] Although American firms were still substantially larger than British firms in 1935, as in 1907, this was now because of the relatively large size of the United States

[3] Leak and Maizels, 'The structure of British industry', p. 157.

[4] If the industrial classification of the 1935 census had been finer, it would have revealed many more cases of industries in which high concentration had been built up by mergers in the preceding decades, see Leak and Maizels, 'The structure of British industry', pp. 163–5.

[5] For a discussion of the methodological issues and the statistics, see: P. S. Florence, *The Logic of British and American Industry* (2nd. ed. 1961), pp. 132, 135; G. Rosenbluth, 'Measures of concentration', in Conference of the Universities – National Bureau Committee for Economic Research, *Business Concentration and Price Policy* (Princeton, 1955), pp. 70–7; W. G. Shepherd, 'A comparison of industrial concentration in the United States and Britain', *Rev. Econ. Stat.*, vol. 43 (1961); J. S. Bain, *International Differences in Industrial Structure* (New Haven, 1966), pp. 76–81; B. P. Pashigian, 'Market concentration in the United States and Great Britain', *Journal of Law and Economics*, vol. 11 (1968). The inference made above does, of course, depend on the assumption that concentration did not decline in the US between 1907 and 1935.

economy, rather than because output in the United States was more highly concentrated in a few firms, as it had been earlier.

This evidence of Britain 'catching up' with the United States between 1907 and 1935 is based on average levels of concentration in individual industries, but the picture of increasing concentration in Britain is confirmed by an independent set of statistics on concentration in manufacturing industry overall. This is simply and conventionally measured as the share of the largest fifty or of the largest 100 firms in the output of manufacturing industry. There is no necessary correspondence between changes in overall concentration in manufacturing, on the one hand, and changes in the share of the top three firms in individual industries, on the other. It would, for example, be possible for the former to rise whilst the latter remained stable, if large firms were growing principally by diversifying over a larger number of industries. However, the infrequency of diversifying mergers suggests that overall concentration measures will not do great violence to the historical facts of concentration, and overall concentration is, moreover, of some interest in its own right.

The most accurate recent estimates of the share of the largest 100 firms in manufacturing output between 1909 and 1970 are presented in Figure 7.1.[6] The points for the years for which data are available are plotted on a log scale graph to facilitate comparison of the rates of growth in the share of the largest 100 firms in each period. All of the estimates, and especially those for the earlier years, are in some degree conjectural. Moreover, the twelve dates pinpointed in the graph are arbitrary in the sense that we are uncertain as to whether the figure for any single year is representative of the surrounding years; and indeed on whether the turning points which appear in the graph are truly turning points, rather than stages on an extended upward or downward movement which would disappear if we had estimates for each year. For these reasons, in interpreting the graph too much emphasis should not be placed on slopes that do not significantly differ from the horizontal: the 'decline' from 1939 to 1948 could, for example, be accounted for entirely by errors in the data.[7] There are, however, some important

[6] These estimates are based on *Census of Production* data and on interpolations by S. J. Prais and others. Because of the nature of the assumptions necessary in the interpolations, they must be regarded as tentative, pending close and direct analysis of the *Census of Production* returns of large firms. See Appendix 2 for a fuller discussion of changes in aggregate concentration over time.

[7] For further discussion, drawing on supplementary evidence, of the (apparently) downward trend of concentration in the 1930s and 1940s, see Chapter 9.

movements in concentration which stand out very clearly in Figure 7.1 and which cannot be written off as resulting from freak data. First, after several decades of perhaps slightly increasing concentration, there was, in the decade following the First World War, a sustained and rapid rise in industrial concentration, as a result of which the largest 100 firms gained control of perhaps one quarter of manufacturing output. There then followed several decades of stagnant, or possibly even declining, concentration so that the level of 1930 was probably not exceeded until the early 1950s, by which time the second substantial upward movement was clearly under way. The rapid rise in the share of the largest 100 firms, with which we are already familiar from postwar evidence, then reasserted itself, and it continued into the 1960s at a rate comparable with that of the 1920s. This second surge has been sustained over a longer period than the previous one, so that by the early 1970s the top 100 firms controlled almost one half of manufacturing output.

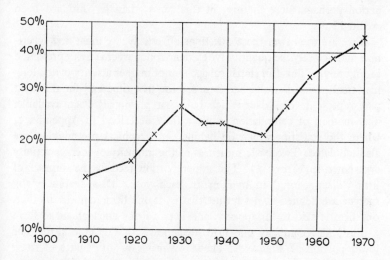

FIGURE 7.1   *The share of the largest 100 firms in manufacturing net output, 1909–1970 (log scale).*

Source: *Appendix 2.*

This chronology provides an important clue to the dynamic process by which the corporate economy developed: a clue revealed by comparison of these changes in concentration with the time pattern of merger activity shown in Figure 7.2 below. Over the period from the turn of the century to the present a broad correlation between merger movements and changes in concentration is evident. In particular the two periods when merger activity was at its most intense – the 1920s and the 1960s – correspond to periods in which the largest 100 firms were increasing their share of output, while the lull in concentration in the 1930s and 1940s is paralleled by two decades of exceptionally low merger activity. Now this correspondence would perhaps not seem remarkable to those unfamiliar with the empirical literature in this field, but previous surveys of mergers have tended to discount the importance of mergers in historical periods and to suggest that only in the 1950s and 1960s did they have a significant impact on industrial concentration.[8] The possibility must be faced that this view, hitherto widely accepted, is misconceived, and from this further questions naturally arise. Were the postwar merger waves really unprecedented, as has been claimed, or were they rather a repetition and extension of something already widely experienced in the 1920s? Was the corporate economy of the inter-war period perhaps more mature, in this sense, than has hitherto been credited?

If we are to answer these questions effectively, we must first scrutinize more closely the quantitative evidence on merger activity presented in Figure 7.2. This full statistical account of merger activity over a long historical period is published here for the first time and for the most part represents new data which have not previously been available. The method of compilation is more fully described in Appendix 1, where the statistics which lie behind our graphical presentation are also tabulated. Two basic measures of the intensity of merger activity are shown in Figure 7.2. The upper portion records the number of firms 'disappearing' in mergers in each year. 'Disappearances' by merger are defined in net terms (that is, if one firm acquires another, one firm is said to 'disappear'; and if two firms amalgamate to form a third, new company one firm 'disappears' in this case also). The lower portion of Figure 7.2 shows estimates of the values of these 'disappearing' firms, corrected for price changes to facilitate comparison between different years. While in individual years the two indicators can show widely differing movements, the broader trends which they show do not markedly differ. Taking the statistics at their

---

[8] E.g. P. E. Hart and S. J. Prais, 'The analysis of business concentration: a statistical approach', *Journal of the Royal Statistical Society*, series A, vol. 119 (1956), pp. 168–9; A. Singh, *Takeovers, their Relevance to the Stock Market and the Theory of the Firm* (Cambridge, 1971), p. 15.

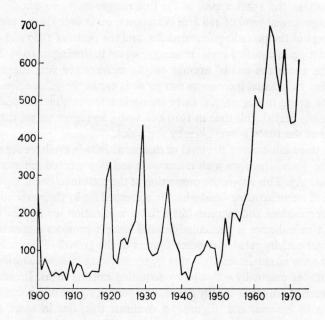

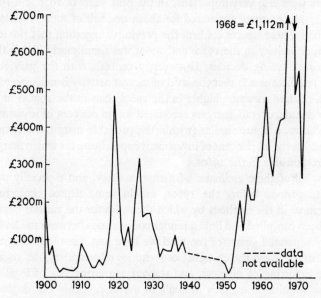

FIGURE 7.2 *Merger activity in UK manufacturing industry, 1900–73.*

face value, the 1920s appear as the first merger intensive decade. The average annual level of 188 firm disappearances is over three times the average of the preceding four decades, and the peaks of 1920 and 1929 are not again equalled until the merger waves beginning in 1959. Then in the 1960s the annual average of 564 mergers per year suggests a further substantial increase in merger activity. In the value series (the lower graph in Figure 7.2) there are again important merger peaks in the early period (this time in 1919 and 1926), and again we see that the level of the 1920s is overtaken by the 1960s.

A third indicator of the level of merger activity is available for some of the years which we wish to compare and is presented separately in Figure 7.3. This shows the proportion of the total investment expenditure of manufacturing firms which is accounted for by their expenditure on acquisitions and mergers (for a fuller explanation see Appendix 1). Such an indicator is particularly useful since it provides a measure of merger activity relative to other sources of the growth of firms. Firms may grow either internally – that is, by investing in new factories and plant – or externally – that is, by acquiring existing firms. Investment in internal growth is usually the most important source of expansion for firms in general, and Figure 7.3 confirms this, but in some years mergers were also very important. In the peak years of 1926 and 1968, for example, mergers accounted for about one half of total investment spending. These figures confirm the previous suggestion that the level of merger activity in the 1920s and 1960s was significantly higher than in the intervening decades. However, a contrast with the previously presented measures is that the level of merger activity is not, according to this relative measure, higher in the 1960s than in the 1920s: in the earlier decade, overall, mergers accounted for 32 per cent of investment expenditure, a figure similar (within the probable margins of error in the data) to the 28 per cent of investment expenditure for which mergers were responsible in the 1960s.

Now all of these estimates of merger activity, and especially those for the periods before the 1960s, are in some degree conjectural. Differences in the methods by which the series for the various decades have been compiled and linked render comparisons between the level in widely separated years (of the kind we have just hazarded) extremely speculative. Given the changing conventions of compilers, the vagaries of the reporting of mergers, and the varying stringency of legal disclosure requirements, it is difficult to assess precisely the degree to which series covering such a long period are comparable. The number of mergers is undoubtedly greater in each period than the level actually recorded in Figures 7.2 and 7.3, but we cannot tell by how much. Moreover, in contrast to the case of the iceberg (whose visible tip always bears the same relationship to its submerged portion), the various peaks seen in the merger statistics may represent a changing proportion of total merger activity, and this would clearly imperil our conclusions. A more detailed analysis of the merger waves at the level of individual firms can, however, be used as an independent check on the impression

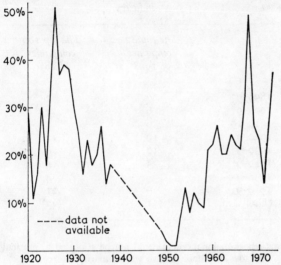

FIGURE 7.3   *Merger values as a proportion of total investment expenditure in UK manufacturing industry, 1900–73.*
Source: *Appendix 1.*

gained from the aggregate merger statistics. We have a very full record of the merger activity of the largest firms because their major mergers are regularly noted in the financial press, and also many of them have published company histories in which we can follow their development. It is therefore possible, for each large firm, to discover in which decade its founding merger or most substantial growth by acquisition occurred. On this basis an intertemporal profile of merger activity can be constructed which will be less subject to the vagaries of reporting that detract from the comparability over time of the aggregate merger statistics. The results of such an inquiry are shown in Table 7.1.[9] The

[9] Some element of subjective judgement enters into the choice of the decade of the major merger: e.g. when a choice has to be made between a founding

right hand column confirms that the impression in the aggregate statistics (of the predominance of the 1920s and 1960s in merger activity) is an accurate one. Of the largest companies in 1970, fourteen had their major merger in the 1920s and as many as twenty-one in the 1960s, both of these figures being well above the number experienced in other decades. There is, however, a tendency for later merger waves to

TABLE 7.1   *The decade of the major mergers of the leading companies in UK manufacturing industry*

	The top 82 companies of 1948	The top 100 companies of 1970
1890–9	7	2
1900–9	6	4
1910–19	6	2
1920–9	19	14
1930–9	10	6
1940–9	3	4
1950–9	–	4
1960–9	–	21
(Unclassified*)	(31)	(43)

– Not applicable.
* Firms whose major mergers could not be allocated to particular decades or whose growth was primarily internal.
Source: see note 9 below.

expunge the effects of previous ones, and this may create a new bias in the results. The GEC, AEI, English Electric merger in 1967–8, for example, is chosen as the major merger for that company, though each of the three component companies also had earlier 'major' mergers which, had they remained independent, would have been recorded in the table. A counterweight to this effect is therefore provided in the left hand column of Table 7.1. This is based on the largest firms in

merger and a subsequent period of substantial growth by acquisition. For an indication of the methods adopted in such cases, see L. Hannah, *The Political Economy of Mergers in Manufacturing Industry in Britain Between the Wars* (unpublished DPhil thesis, Oxford, 1972), pp. 143–4, 272–8. The companies of 1948 are the largest manufacturing companies (ranked by net assets) in the National Institute of Economic and Social Research list for that year. The companies of 1970 are the largest (ranked by sales) in the *Times 1000* list.

1948, and thus shows the role of mergers in the 1920s more clearly, without the overlay of subsequent merger waves.[10] The table as a whole leaves little room for doubt that there was a substantial merger wave in the 1920s which perhaps rivalled in importance that of the 1960s. The contrast between this and the experience of the United States is particularly striking. A similar study of the largest 100 US manufacturing corporations showed that over half of their major mergers occurred before the First World War.[11] Thus the picture of British firms 'catching up' in the period around the 1920s with a structure of industry which had been created earlier in the United States is again confirmed.

We are thus faced with a range of independent sets of statistics which suggest that the period from the closing years of the First World War to the early 1930s saw a significant change in the structure of British manufacturing industry.[12] Large firms were consolidating their hold on the economy, raising their share of manufacturing output by perhaps nine percentage points from 17 per cent in 1919 to 26 per cent in 1930, a rate not exceeded since. At the same time all of the statistical series suggest that the merger activity of the 1920s was intense, both in the aggregate and in the history of individual large corporations. Despite reasonable doubts that may exist about any single set of the statistics presented,[13] the impressions which they collectively support are difficult to refute. It seems reasonable to infer that increasing industrial concentration was the result of the intense merger wave of the 1920s with which the expansion of large firms coincided. But can this inference be confirmed by a more detailed analysis of the impact of mergers on concentration?

Ideally we should like to examine the total impact of merger activity

[10] The results for the large firms of 1948 are of course also different from those of 1970 because of a substantial turnover among large firms between those dates; see G. Whittington, 'Changes in the top 100 quoted companies in the United Kingdom 1948 to 1968', *Journal of Industrial Economics*, vol. 21 (1972).

[11] R. L. Nelson, *Merger Movements in American Industry 1895–1956* (Princeton, 1959), p. 4. Thirty-seven of Nelson's 100 corporations had no important mergers, the others registering eleven before 1895, twenty in 1895–1904, seven in 1905–14, five in 1915–24, eleven in 1925–34, none in 1935–44, and nine in 1945–55.

[12] There were also significant mergers in this period which fall outside the purview of the present study (which is confined to manufacturing industries), e.g. the regrouping of railways and further consolidation of road passenger transport undertakings, cinema chains, petrol distribution, and the financial and retail sectors.

[13] These doubts are more fully discussed in Appendices 1 and 2.

over the whole range of firms between, say, 1919 and 1930. Unfortunately this is not possible because of the lack of data on the sizes of most firms, but we can approximate the sizes of a large proportion of them. Drawing on a recent study,[14] which estimated the sizes of the largest firms (principally quoted companies) in 1919 and 1930, we can assess the contribution of mergers to increasing concentration within this population. Because the great majority of large firms, accounting for perhaps one half of manufacturing industry profits, are included in this population, we can reasonably expect that changes within it will accurately reflect the important structural changes in the corporate economy of these years. The results of the analysis of this population are summarized in Table 7.2. Comparison of the first with the second

TABLE 7.2   *The contribution of mergers to increases in concentration, 1919–30*

	1 *Actual population in January 1919*	*2* *Actual population in December 1930*	*3* *Counter-factual 'merged' population*	*4* *Proportion of the change due to mergers*
Number of firms	1263	584	592	99%
Share of the largest 5 firms	12·6%	27·6%	21·3%	58%
Share of the largest 50 firms	43·4%	63·7%	57·3%	68%
Share of the largest 100 firms	56·4%	77·4%	72·5%	77%

Source: Calculated from original data described in Hannah and Kay, *Concentration in Modern Industry*.

column of the table confirms the strong upward trend of concentration: the number of firms in the population is halved, the share of the top five firms more than doubles, and the share of the top fifty and 100 firms both increase markedly. In order to assess the contribution of mergers to this increase, it is necessary to make an assumption about what would have happened to concentration in the absence of merger

[14] L. Hannah and J. A. Kay, *Concentration in Modern Industry: Theory, Measurement and the UK Experience* (forthcoming 1976). The measure of size adopted was the market value of the firm's capital.

activity. Suppose that in 1919 there were two firms, A and B, each with one half of the market. If A acquired B early in 1919, thus becoming the only firm, and if by 1930 the market had grown by 50 per cent, two views might be taken about what had caused the increase in concentration. One view might be that the merger of 1919 accounted for an initial increase but that the rest of the increase in A's size by 1930 should be accounted internal growth. An alternative, and perhaps more plausible,[15] view would be that it was the merger of A and B that caused the whole of the increase in concentration (since if that merger had not occurred B would have shared the subsequent growth of the market, and the shares of the two companies in the larger market of 1930 would have been the same as in 1919). It is this assumption that is adopted in calculating the percentage contribution of mergers to the increase in concentration which is shown in the fourth column of Table 7.2. This calculation requires the construction of a hypothetical population, reflecting an adjustment to the real firms of 1919 to show them as they would have been if all the mergers which occurred between 1919 and 1930 had occurred instantaneously, in 1919. By comparing the level of concentration in this hypothetical or 'counterfactual' population (column 3) with the level actually observed in 1919 and 1930 (columns 1 and 2), we can estimate the proportion of the increase which is attributable to merger activity. The percentage contribution (column 4) is in all cases found to be above 50 per cent, suggesting that mergers were the most significant cause of increasing concentration in this period. This conclusion is particularly strong since only a part of the total merger activity is included in the hypothetical population: if all of the acquisitions of small firms by large ones had also been included, the dominant contribution of merger activity would have shown up even more clearly. This general conclusion on the importance of mergers in the growth of the larger firms is also confirmed by more comprehensive measures of concentration and by the results for individual industries. These results (which are more fully described in Appendix 2) show substantial increases in concentration in each of fifteen broad industrial groups, and, in all except the vehicles and clothing groups, mergers appear to cause the greater part of the increase in concentration.[16]

[15] For a persuasive argument to this effect, see G. J. Stigler, 'The statistics of monopoly and merger', *Journal of Political Economy*, vol. 64 (1956).

[16] An earlier study by P. E. Hart and S. J. Prais ('The analysis of business concentration: a statistical approach') reached the contrary conclusion that mergers were quite unimportant as a source of increasing concentration

A number of the important dimensions of the rise of the corporate economy do, then, support the conclusion that, in the 1920s, concentration was rapidly increasing as a result of a merger wave of unprecedented proportions and of permanent significance. What is thus revealed in the statistics is broadly consonant with the shifts of opinion amongst businessmen, financiers and politicians which we have chronicled in earlier chapters. The heightened faith in 'rationalization' generally, and in the achievement of economies of scale and coordination in large combines particularly, was clearly being acted upon in these years. The disadvantages of the inherited structure had been perceived and there was an attempt, showing clearly in rising concentration and merger activity, to remedy some of its deficiencies. Processes of this kind are, we began this chapter by conceding, of necessity gradual. Sometimes, it is true, they are hurried on by waves of fashion, but usually they require many years of trial and error, of learning and teaching, before their effects become pervasive and irresistible. To attempt to identify 'stages' in such gradual, evolutionary processes is perhaps arbitrary, but in this case we have good reason for settling a label on one important 'stage'. The years after the First World War clearly saw an acceleration of structural change and it is to the interwar decades, and particularly to the period 1919–30, that the 'rise of the corporate economy' is most naturally dated.

These interwar decades have in general been painted by historians in melancholy shades; and that the underlying depression in the staple industries and the widespread unemployment of able bodied adults represented a blight on industrial life, a deeply unacceptable face of capitalism, can hardly be gainsaid. It has, however, become increasingly clear that against this gloomy backcloth Britain witnessed in the interwar period a rate of economic growth perhaps somewhat higher than that of some major competitors, and certainly better than the growth rate that had been achieved in the decade of stagnation before the First World War.[17] This change in trend growth performance clearly required rapid and substantial readjustments in the structure of in-

---

between 1885 and 1950; but their results are invalidated both by the use of an inappropriate measure of concentration and by the omission from their study of 99 per cent of merger activity, including four-fifths of the largest mergers of the interwar period. Nonetheless, in the absence of better evidence their results have been widely quoted; see, e.g., M. A. Utton, *Industrial Concentration* (Harmondsworth, 1970), p. 88.

[17] J. A. Dowie, 'The interwar period: some more arithmetic', *Economic History Review*, vol. 21 (1968).

dustrial output,[18] and the question naturally arises as to whether the evolving structure of firms which we have described was an integral part of this shift. Economic growth, involving the expansion of new industries and the contraction of old ones, can in principle be achieved by the multiplication of new firms coupled with the bankruptcy of firms in the old industries. However, the changes in the composition of output between the wars did not occur exclusively in this way but rather evolved in parallel with substantial internal and external growth by large firms. Hence it is possible that the merger waves and the growth of large firms were integral parts of a common process of economic change. Contemporary rationalizers, of course, saw the causation as running from structural change to economic growth, and we shall return to this hypothesized relationship between growth and industrial concentration in the final chapter. The causation may, however, also have run in the opposite direction; the greater the changes in the structure of output towards more inherently concentrated and rapidly growing industries, the higher will be the overall level of concentration. In the next chapter, therefore, we must examine more systematically the course which the large corporations were steering: the directions of their expansion, and in particular their role in the new industries.

[18] H. W. Richardson, 'The new industries between the wars', *Oxford Economic Papers*, vol. 13 (1961).

8

The rise of the corporate economy: directions

It may be that the influence of Mass Production upon the
Destinies of Great Britain will be far greater than
that of the Wars and adventures from which
the British Empire grew.

SIR ERIC GEDDES, *Mass Production*
(1931), p. 5.

ಇಇ

The industries that by the 1930s were the most highly concentrated
were, in general, also those that had grown most rapidly in the previous
two decades. This is true both of the totally 'new' industries such as
gramophones or rayon (which had not existed at the turn of the century) and also of broader industrial groups, like chemicals and electrical
engineering, which had been growing faster than the average.[1] The
seven industrial groups which had the highest average three-firm concentration ratios in 1935, for example, accounted for 46 per cent of
industrial output in 1907, but for as much as 56 per cent of the higher
industrial output of 1935.[2] This suggests that the overall rise in concentration which we have noted is due in part to the growing importance
in the economy of 'new', fast growing and perhaps inherently more
concentrated industries.[3] However, this can only explain part of the
increase, for there were also substantial changes in concentration
within individual industries. Many of the concentrated industries of
1935 had achieved that position in the 1920s and had been relatively

---

[1] H. W. Richardson 'The new industries in Britain between the wars', *Oxford
Economic Papers*, vol. 13 (1961), reprinted in D. H. Aldcroft and H. W.
Richardson, *The British Economy 1870–1939* (1969). H. Leak and A. Maizels,
'The structure of British industry', *Journal of the Royal Statistical Society*,
series A, vol. 108 (1945).

[2] Calculated from information in: Leak and Maizels, 'The structure of British
industry', p. 157; B. R. Mitchell and P. Deane, *Abstract of British Historical
Statistics* (Cambridge, 1962), p. 270.

[3] A. F. Lucas, *Industrial Reconstruction and the Control of Competition* (1937),
p. 40.

unconcentrated in 1907.[4] Indeed, the share of the largest five firms is known to have risen between 1919 and 1930 in every one of the fifteen broad industry groups, mainly as a result of the substantial waves of merger activity in the 1920s (see Appendix 2).

The role both of the rise of new, concentrated industries, and of increasing concentration within industries, can be seen clearly in the changing composition of the dominant enterprises in the economy (which would also tend to be those which dominated individual industries). Tables 8.1 and 8.2 list the largest fifty companies[5] (by market valuation) in January 1919 and twelve years later, when the main effects of the merger wave had been felt, in December 1930.[6] As one would expect from the overall increase in concentration, the largest fifty in 1930 were considerably larger than those of 1919.[7] A noticeable feature of the tables is that only one half of the top fifty firms of 1919 are still in the group of the largest firms twelve years later in 1930.[8] The principal reason for this is that many of them, notably the heirs of the turn of the century merger wave, were in industries like textile finishing which faced declining markets in the interwar years. Firms like Armstrong-Whitworth, the naval shipbuilders, and the Bradford Dyers Association, though they had been prosperous before

[4] E.g. cast iron and steel pipes, explosives, matches, polishes, primary batteries, soap, spirit distilling, sugar, weighing machinery, wrought iron and steel tubes. All of these industries had three-firm concentration ratios in excess of 70 per cent by 1935 and had experienced substantial merger activity in the previous decades.

[5] Oil companies like Shell and British Petroleum are excluded from these lists since, although they were large, they did not have substantial manufacturing interests in the UK at this time, but were involved mainly in overseas operations and in UK marketing.

[6] The market values of quoted capital were calculated from data in the *Stock Exchange Daily Official List* and estimates were made for the unquoted capital using data on nominal capital, dividend yields, etc. These approximations provide only rough order of magnitude estimates of their sizes, and in individual cases may be misleading. For example, Ford appears in the table to be larger than Morris Motors, though in terms of UK output it was the smaller firm. This was because Ford's capital included significant European assets and also because Morris's unquoted ordinary share capital has probably been undervalued. In general, however, the tables serve well their major purpose of identifying the larger corporations.

[7] Share prices in 1930 were only 15 per cent higher than in 1919, whilst the values were 125 per cent higher, so the change is a real one, and not merely the consequence of inflation.

[8] The precise number of survivors is a question with no clear answer, since many 'survived' only after being consolidated into large corporations. The number of corporations whose *names* survived is somewhat less than half because of such new consolidations.

the war, were not destined to become the basis of the modern corporate economy. Thus fourteen of the largest fifty firms in 1919 not only failed to hold their own relative to other firms but showed a decline in absolute terms also: in 1930 the average value of their capital was less than two-thirds of its value in 1919. Though some of these firms were eventually to recover from the depression, many of them showed little further growth and clearly belonged more to the nineteenth century than to the modern corporate economy.

TABLE 8.1   *The fifty largest companies of 1919*

Industry	Name of company	Estimated market value (£million)
Textiles	J. & P. Coats	45·0
Food	Lever Bros	23·0
Tobacco	Imperial Tobacco	22·9
Shipbuilding/engineering	Vickers	19·5
Drink	Guinness	19·0
Chemicals	Brunner Mond	18·7
Chemicals	Nobel Industries (Explosives Trades)	16·3
Textiles (Rayon)	Courtaulds	16·0
Vehicles	Metropolitan Carriage Wagon and Finance	14·4
Metal manufacture	United Steel	13·2
Shipbuilding	Armstrong-Whitworth	12·2
Textiles	Fine Spinners & Doublers	9·9
Building materials	Associated Portland Cement Manufacturers	9·1
Rubber	Dunlop Rubber	8·9
Chemicals	Reckitt & Sons	8·8
Metal manufacture	Guest Keen & Nettlefold	8·2
Chemicals	Levinsteins (British Dyestuffs Corporation)	8·1
Shipbuilding	John Brown	7·7
Drink	Watney Combe Reid	6·9
Textiles	Bleachers Association	6·7
Metal manufacture	Consett Iron	6·6
Engineering	Babcock & Wilcox	6·5
Metal manufacture	Dorman Long	6·5
Textiles	Calico Printers Association	6·4
Food	Maypole Dairy	6·2
Metal manufacture	Richard Thomas	6·2

TABLE 8.1 (*continued*)

Industry	Name of company	Estimated market value (£million)
Textiles	Bradford Dyers Association	6·1
Chemicals	United Alkali	6·1
Publishing	E. Hulton & Co.	6·0
Metal manufacture	Mond Nickel	5·5
Drink	Bass Ratcliffe Gretton	5·3
Engineering	Hadfields	5·3
Drink	Buchanan-Dewar	5·2
Chemicals	Boots Pure Drug	5·0
Textiles	English Sewing Cotton	5·0
Metal manufacture	Stewarts & Lloyds	5·0
Shipbuilding	Cammell Laird	4·8
Engineering	Birmingham Small Arms	4·6
Textiles	Horrockses Crewdson	4·5
Chemicals	Borax Consolidated	4·4
Textiles	Linen Thread Co.	4·4
Drink	Distillers Company	4·3
Food	Liebigs Extract of Meat	4·3
Food	J. Lyons	4·3
Publishing	Amalgamated Press	4·2
Publishing	Associated Newspapers–Daily Mail Trust	4·2
Metal manufacture	Ebbw Vale	4·1
Engineering	Platt Bros	4·0
Metal manufacture	John Lysaght	3·9
Electrical engineering	English Electric	3·8

Source: See p. 117, note 6.

TABLE 8.2   *The fifty largest companies of 1930*

Industry	Name of company	Estimated market value (£million)
Food	Unilever	132·0
Tobacco	Imperial Tobacco	130·5
Chemicals	Imperial Chemical Industries	77·3
Textiles (Rayon)	Courtaulds	51·9
Textiles	J. & P. Coats	47·4
Drink	Distillers Company	45·5
Drink	Guinness	43·0
Rubber	Dunlop Rubber	28·2
Publishing	Allied Newspapers	27·6
Vehicles	Ford Motor Company (British subsidiary of U.S. Ford)	21·2
Metal manufacture/ engineering	Guest Keen & Nettlefold	20·3
Shipbuilding/engineering	Vickers	19·6
Drink	Watney Combe Reid	18·5
Publishing	Daily Mail Trust	15·6
Electical engineering	General Electric Company	14·5
Building materials	Associated Portland Cement Manufacturers	13·9
Chemicals	Reckitt & Sons	13·4
Drink	Bass Ratcliffe Gretton	13·3
Chemicals	Boots Pure Drug (British subsidiary of Liggett International)	12·8
Asbestos	Turner & Newall	12·6
Food	J. Lyons	12·1
Engineering	Babcock & Wilcox	11·0
Food	Bovril	10·8
Food	British Cocoa & Chocolate (Cadbury-Fry)	10·3
Metal manufacture	Mond Nickel (British subsidiary of International Nickel)	10·3
Tobacco	Carreras	10·0
Drink	Walker-Cain	10·0
Drink	Mitchells & Butler	9·9
Electrical engineering	Associated Electrical Industries (British subsidiary of International General Electric)	9·8

TABLE 8.2 (*continued*)

Industry	Name of company	Estimated market value (£million)
Metal manufacture	Dorman Long	9·8
Textiles	Fine Spinners & Doublers	9·8
Food	Tate & Lyle	9·3
Publishing	Daily Mirror Newspapers	8·8
Clothing and footwear	J. Sears (Trueform Boot Co.)	8·5
Matches	British Match Corporation	8·1
Paper	Wallpaper Manufacturers	7·9
Drink	Hoare & Co.	7·6
Vehicles	Morris Motors	7·6
Electrical engineering	British Insulated Cables	7·5
Drink	Taylor Walker	7·5
Food	United Dairies	7·3
Drink	Barclay Perkins	7·1
Metal manufacture	Stewarts & Lloyds	7·1
Textiles	Bleachers Association	6·9
Food	Liebigs Extract of Meat	6·9
Textiles (Rayon)	British Celanese	6·8
Metal manufacture	United Steel	6·7
Drink	Allsopp	6·4
Textiles	Combined Egyptian Mills	6·3
Chemicals	Pinchin Johnson	6·3

Source: see p. 117, note 6.

Yet such a fate was not inevitable, for among the more prosperous survivors of the 1919 group there were ten companies which, though based in nineteenth century industry, were sufficiently large, or sufficiently resourceful in the new interwar environment, to maintain their position. J. & P. Coats, Fine Spinners & Doublers, and Stewarts & Lloyds, for example, maintained their size, while GKN and Imperial Tobacco managed a considerable expansion. Other firms, like Vickers, though faced with drastically declining markets, contrived to maintain their size by a programme of diversification and acquisition. Associated Portland Cement Manufacturers, by contrast, grew almost entirely on the basis of the industry which it had dominated since its formation in 1901, and, by investing in large rotary kilns and acquiring rivals, it consolidated its market position and shared in the general interwar prosperity of the building trades.

It is the growth of firms in the new industries, however, that is most striking both in terms of its immediate quantitative impact seen in Table 8.2 and for its strategic significance in the corporate economy. Rapidly growing industries such as rayon, motor cars, electrical engineering and the modern sectors of the chemical trade experienced significant changes in the structure of their firms. Before the First World War, the British economy had been overcommitted to the old industries, and, for reasons which are only just beginning to be systematically investigated by historians, there had been a failure to invest in the new and potentially rapidly expanding sectors, so that for products like dyestuffs and motor cars imports had exceeded domestic production.[9] In the First World War and after, however, a dramatic shift in the pattern of world trade, and a realization of the potential profitability of a pattern of production more akin to that in America and Germany, seems to have induced a substantial shift; and in the eleven years 1920–30 one third of gross capital formation was directed to five major growth industries: rayon, electricals, motors, chemicals, and paper.[10] In many of these industries, therefore, the growth of firms was rapid and as the new capacity often incorporated mass production techniques it also tended to be concentrated in relatively few firms. Thus William Morris was able, by reaping economies of scale, to expand his modest 5 per cent share of the motor car market in 1919 to 41 per cent by 1925, increasing his annual production from only 387 cars to 55,582 in the same period. Other car manufacturers, notably Ford (a subsidiary of US Ford) and Austin, also grew substantially in the 1920s, as did the major domestic supplier of motor car tyres, Dunlop Rubber.[11] In the rayon industry, in which Courtaulds had already attained a dominant position in 1919, there was also substantial further growth, and by the later 1920s the company was providing a substantial proportion of the world's greatly enlarged output. At the same time its major competitor in the rayon industry, British Celanese, was also growing rapidly and joined the list

9 See, generally: H. W. Richardson, 'Overcommitment in Britain before 1930', *Oxford Economic Papers*, vol. 17 (1965), reprinted in Aldcroft and Richardson, *The British Economy 1870–1939*; W. P. Kennedy, 'Foreign investment, trade and growth in the United Kingdom, 1870–1913', *Exploration in Economic History*, vol. 11, 1974).

10 H. Tyszynski, 'World trade in manufactured commodities, 1899–1950', *Manchester School*, vol. 19 (1951). D. H. Aldcroft, 'Economic progress in Britain in the 1920s', *Scottish Journal of Political Economy*, vol. 13 (1966), reprinted in Aldcroft and Richardson, *The British Economy 1870–1939*.

11 P. W. S. Andrews and E. Brunner, *The Life of Lord Nuffield* (Oxford, 1955), p. 112. G. Maxcy and A. Silberston, *The Motor Industry* (1959), pp. 13–16.

of the fifty largest firms.[12] The mass production methods introduced by these firms enabled them to make substantial price reductions. The average factory price of cars fell from £308 in 1912 to £259 in 1924, £206 in 1930, and £130 by 1935–6, while the price of a hank of viscose yarn fell from a postwar peak of 19s. 3d. to 4s. 6d. in 1929, and 2s. 6d. in the later 1930s.[13]

The innovation of mass production methods, which was so clear a feature of the manufacture of rayon and motor cars, was less readily apparent in the science based, and rapidly expanding, electrical engineering and chemical industries, but here also the size of firms greatly increased. Whilst there were still opportunities for small, specialist firms in many branches of these industries, a number of large corporations developed, with widespread interests diversified over a range of manufactures. The formation of such companies appears to have owed a lot to the merger waves between the closing years of the First World War and the early 1930s. Imperial Chemical Industries, for example, was formed in 1926 as a consolidation of four of the largest fifty firms of 1919: Nobel Industries, Brunner Mond, the British Dyestuffs Corporation, and the United Alkali Company.[14] All except the latter were themselves recent groupings of chemical companies – Nobel was a merger of thirty explosives companies in 1918 and had since made a number of further acquisitions in an attempt to diversify. Brunner Mond, already large before the war, had acquired a number of rivals and integrated backwards during the war. British Dyestuffs, which controlled 75 per cent of British dyestuffs output, was the result of a merger backed by the government during the war in an effort to expand the British production of dyestuffs. After the main ICI merger this pattern of consolidation and diversification by merger continued, and further significant purchases were made of firms in the fertilizer, metals and dyestuffs fields.

[12] J. Harrop, 'The growth of the rayon industry between the wars', *Yorkshire Bulletin of Economic and Social Research*, vol. 20 (1968). D. C. Coleman, *Courtaulds, an Economic and Social History*, vol. 2 (Oxford, 1969), pp. 171–204.

[13] D. H. Aldcroft, *The Interwar Economy: Britain 1919–1939* (1970), pp. 184, 189.

[14] The comments on ICI which follow are based on an examination of the company's archives, now held by the ICI parent company. The files of the Technical Department and the Development Department were particularly useful. See also: W. J. Reader, *Imperial Chemical Industries: A History*, vol. 1 (1970), pp. 249–466; vol. 2 (1975); J. Stamp, 'Amalgamations', in *Some Economic Factors in Modern Life* (1929).

The advantages of such a grouping are to be seen partly in the classical motives of the reduction of competition and the achievement of scale economies. Competition was reduced not only by the increased market dominance of ICI in areas such as dyestuffs, but also by forestalling the possibility of new entry by one of the other merger partners which did not at that time directly compete but which had the technical ability to diversify competitively in the future. Economies of scale were achieved partly by the rationalization of old plant and partly through new investment in larger plant. Six months after the initial merger, for example, Nobel had closed 55 per cent of its explosives capacity and was producing for the lower postwar demand at a cost reduction of 16 per cent. Similar rationalization appears to have followed the main ICI merger: the alkali production of the United Alkali Company and Brunner Mond was, for example, concentrated in the more efficient plants, with resulting cost savings overall. Such concentration could, in principle, have been achieved without a merger, by competitive forces (as in the motor car industry), but in practice competition in these branches of the chemical industry was so imperfect that in many cases merger was probably a prerequisite of such cost savings and certainly speeded up the process. New investment projects on a large scale were also undertaken by the new ICI group: the Billingham complex, for example (which Brunner Mond had had difficulty in financing alone), was significantly expanded. Perhaps the most significant contribution of the merger, however, was in bringing together the commercial and technical expertise of firms in heavy chemicals, organic chemicals, chemical engineering and non-ferrous metals in a unique complementary combination. It had been the fear of Nobel (which brought its expertise in nitrocellulose and non-ferrous metal technology to the group) of 'having no technical and economic groundwork on which to build'[15] that had been one of the original motives of the ICI merger. With the benefit of the merger, however, the diversified expertise of the constituent firms was applied to a wide range of products, including paints, metals, leathercloth, plastics, solvents, dyestuffs, fertilizers, and high pressure engineering, and the collaboration was generally a fruitful one. Whilst not all the resulting spin-offs and new investments were profitable, many were, and the grouping made a significant contribution to the diversification and modernization of the mix of output in the chemical industry in which ICI held a dominant share.

[15] L. J. Barley, 'Memorandum on development policy', dated 27 Apr. 1926 (Nobel archives).

In the electrical engineering industry concentration was less marked,[16] but three principal groupings came to fill a role similar to that of ICI in the chemical industry. The largest of these – the General Electric Company – had its origins in an electrical merchanting business of the 1880s and, after substantial prewar growth, embarked upon an important programme of expansion in the war, taking control of the Osram Company (which had been cut off by the war from its German parent), and Chamberlain & Hookham (a meter manufacturer). However, the claim of the company to manufacture 'Everything Electrical' was only really assured with the acquisition during the war of Fraser & Chalmers, which secured a place in the heavy electrical industry. The company's Witton works were also extended and a 50 per cent interest was taken in Pirelli-General Cables. Between 1918 and 1922 GEC's issued capital was increased from £1½ million to almost £9 million, so that the stage was set for the company's expansion from a work force of 15,000 in 1919 to 40,000 twenty years later. Somewhat less successful was English Electric, another wartime grouping of electrical companies manufacturing a range of specialist electrical products, based on Dick Kerr & Company of Preston. After purchasing the Coventry Ordnance Works from a consortium of shipbuilders, this group was floated as the English Electric Company, with a capital of £5 million, in 1919, in which year it also acquired the former Siemens Dynamo Works at Stafford for £1 million. Although the company subsequently experienced serious financial and managerial problems (and hence was not among the largest fifty firms of 1930) it became a major rival for GEC after reorganization in 1931.

The origins of the third leading electrical manufacturer, Associated Electrical Industries, are bound up with American influences. The British Thomson-Houston Company was already owned by the General Electric Company of America (no relation to the British GEC), and Gerard Swope, the president of the American parent company, was anxious to extend his influence in Britain. Hence in 1926–7 he acquired Ferguson Pailin and Edison Swan and in 1928 added the Metropolitan-Vickers Company (which, though originally under the control of the American Westinghouse Company, had passed from them to Vickers

---

[16] The three leading electrical engineering companies' histories are described in R. Jones and O. Marriott, *Anatomy of a Merger: A History of GEC, AEI, and English Electric* (1970), on which the following two paragraphs draw substantially. See also: Monopolies and Restrictive Practices Commission, *Report on the Supply of Electric Lamps* (1951), and *Report on the Supply and Exports of Electrical and Allied Machinery and Plant* (1957).

during the war), consolidating all four British subsidiaries under the name of Associated Electrical Industries.[17] As a result of this merger activity, GEC, English Electric and AEI came to control some 35 per cent of the electrical engineering industry, and, on the heavy plant side (where fixed costs were large and competition fierce during the periodic depressions in demand),[18] their market share was as high as 60 per cent. The motive of greater market control and the control of competition was certainly an important one in the mergers, and whilst some economies of operation were possible, integration within the merged companies was very slow. In AEI, for example, the BTH and Metrovick interests were run as virtually independent subsidiaries on the manufacturing side for many decades, with only loose coordination of commercial and financial policy. There were benefits in the mergers from the bringing together of commercial manufacturing and technical experts in the electrical industry,[19] but despite this advantage there remained great scope for the smaller specialist firm in many areas. In domestic electrical appliances, for example, production expanded rapidly in the 1930s, but subsidiaries of foreign firms such as Hoover and Electrolux and new British enterprises such as Morphy Richards played the leading role.[20] In the field of electrical sound and entertainment, however, a large British company, Electrical & Musical Industries (a merger in 1931 of the Gramophone Company and Columbia Graphophone), was a conspicuous success.

What both ICI and the leading electrical manufacturing companies had in common was a commitment to, and a developed expertise in, a specific, but wide, area of modern technology. Such specialization was a logical extension of the division of labour in an increasingly complex industrial environment. Moreover, since in many of their individual product lines they faced competition from smaller rivals, we may

---

[17] The company was American controlled until 1934 when British shareholders regained majority control; see Jones and Marriott, *Anatomy of a Merger*, p. 108; cf. Monopolies and Restrictive Practices Commission, *Report on the Supply and Exports of Electrical and Allied Machinery and Plant*, p. 102.

[18] PRO/BT/55/49. P. Cunliffe-Lister, memorandum dated 22 November 1927, p. 2. Cf. G. B. Richardson, *The Future of the Heavy Electrical Plant Industry* (1969).

[19] Though this did not show up in the productivity record of the industry, which was among the worst of the new industries, perhaps because the manufacture of large generating plant to individual specifications did not allow much scope for mass production techniques; see K. S. Lomax, 'Growth and productivity in the United Kingdom', *Productivity Measurement Review*, vol. 38 (1964), p. 21.

[20] T. A. B. Corley, *Domestic Electrical Appliances* (1966), pp. 36–48.

perhaps infer from their continued growth[21] that there were real advantages to the possession and development within the firm of such a core of technical expertise.[22] This strategy of development could also be successful in narrower technical fields. Dunlop, for example, achieved substantial growth in the rubber industry not only on the basis of the greatly increased interwar demand for its traditional product, motor car tyres, but also by integrating backwards and forwards and by diversifying within the rubber industry. Hence it became involved in rubber plantations, in the production of cotton and wheels (both related to its tyre production) and in the marketing of tyres and retreading services, and also acquired the Charles MacIntosh group and other companies in the general rubber goods trade.[23] Similarly, Turner & Newall, initially a merger in 1920 of four manufacturers involved in the asbestos trade, pursued a policy of diversification and integration, acquiring in 1925–30 Ferodo brake linings, a group of asbestos cement companies, and a substantial interest in Rhodesian asbestos mines.[24]

Many of these technically based firms in the new and expanding industries were also developing sources of future growth by heavy investment in research and development.[25] There are significant economies of scale and advantages to the diversification of risks in corporate research, and it was generally felt that a firm employing less than 1000 people was unlikely to mount a significant research effort. Hence the growth of larger firms did much to stimulate this kind of expenditure, and, in consequence, accelerated the rate of technical

[21] But cf. pp. 144–5 below for some indication that in the 1930s and 1940s the growth of large firms was relatively unimpressive.

[22] See, generally, G. B. Richardson, 'The organisation of industry', *Economic Journal*, vol. 82 (1972), for a formal treatment of the reasons for the specialization of firms by capabilities. See also pp. 72–3 above for the advantages of a core of technical expertise in appraising investment projects.

[23] P. Jennings, *Dunlopera* (privately published, 1961), pp. 131–2. Monopolies and Restrictive Practices Commission, *Report on the Supply and Export of Pneumatic Tyres* (1955) and *Report on the Supply of Certain Rubber Footwear* (1956).

[24] Monopolies Commission, *Report on the Supply of Asbestos and Certain Asbestos Products* (1973). N. A. Morling, 'History of Turner Brothers Asbestos Co. Ltd', *Rochdale Literary and Scientific Society Transactions*, vol. 24 (1961).

[25] The paragraphs which follow draw extensively on: M. Sanderson, 'Research and the firm in British industry 1919–1939', *Science Studies*, vol. 2 (1972); L. F. Haber, 'The British chemical industry between the wars' (unpublished, 1972); L. Hannah, 'Applied science and research expenditure in twentieth century Britain' (unpublished, 1972).

innovation. The constituent firms of ICI, for example, had all financed research laboratories, but within four years of the merger of 1926, their collective research expenditure was, as a matter of deliberate policy, quadrupled to reach £1 million by 1930, and, after a cutback in the depression, it rose again to £1·4 million by 1939. This represented perhaps three-quarters of the total research effort of the chemical industry and extended over a wide field of chemical products from fertilizers and dyestuffs to the high pressure chemical engineering technology of Billingham. Similar expansion took place in other industries. GEC, which had set up a research group just after the First World War, built its large Wembley laboratories in 1923, and by 1927 was employing 200 research staff there. British Celanese, Courtaulds' major rival in the rayon industry, revealed in 1932 that in the previous seven years it had spent over £1 million on research into the acetate process. Many others among the leading firms also established research laboratories and even in the older established industries firms began investing more in research than had been common in the past. The United Steel Companies, for example, opened their large central research department at Stocksbridge in 1934.

In 1926, when Harry McGowan of Nobel had speculated on the future of the chemical industry, he had felt that 'organized research lies at the root of our prosperity', and he referred explicitly to the need to copy the German example.[26] Seven years later, when he had implemented this policy within ICI, he was well satisfied with the results: '[The] benefits [of research] are continuous. They may be summarized as more economical manufacturing processes, improved outputs, finer products, more efficient technical services . . . and the development of new commodities.'[27]

These favourable sentiments were also echoed by Lord Rutherford, in his last report as Chairman of the Department of Scientific and Industrial Research:

> The historian of the future [he wrote in the mid-1930s] will probably point to the last five years as a period marking an important development in the industrial outlook of this country. These years have seen the fruition of the policy adopted by several large undertakings of setting well-balanced teams of research workers to solve a particular problem or to develop a new product. . . .

[26] Reader, *Imperial Chemical Industries*, vol. 1, p. 414.
[27] *Economist* (15 Apr. 1933), p. 830.

Cooperation, team work and an extensive organization on the technical side are essential for success.[28]

Out of the investment in corporate laboratories came new products such as television and plastics, and improvements in fuel utilization and electric lamps; and the success of such innovations had a strong bearing on the commercial survival and growth of the inventive firms concerned. Patent applications from companies (as opposed to individuals), which had accounted for only 15 per cent of the total before the First World War, had risen to 58 per cent by 1938, and a study of individual major inventions also shows the increased role played by industrial laboratories both in the original invention and in the development and marketing of new products.[29] Among the many 'springs of technical progress' which improved the productivity performance of British firms in the interwar years, then, organized research in corporations with a specialist core of technical knowledge was certainly one.[30]

In addition to new products and new techniques based on scientific developments, there were also opportunities for corporate growth based on specialist skills in marketing. This is reflected in the growing number of large enterprises which were based in the consumer goods market. Firms that had already developed skills in advertising in the nineteenth century, like Guinness, continued to show a high rate of expansion, and further consolidation of other nationally advertised brands of alcoholic drinks occurred: Bass, for example, acquired Worthington for £3 million in 1927, and the Distillers Company integrated forward to acquire the two major blended whisky groups, Johnnie Walker and Buchanan-Dewar, for £20 million in 1925. The foremost exemplars of corporate growth based on marketing were, however, in the food industry. The importance of branded goods and advertising was well established by the First World War, and the forward integration of manufacturers into distribution and wholesaling (which branding often entailed) was given an additional boost by the increasing competi-

---

[28] Department of Scientific and Industrial Research, *Report of the Year ending 31 March 1936* (Cmd. 5350, 1937), p. 15.

[29] J. Jewkes, D. Sawers and R. Stillerman, *The Sources of Invention* (1958), p. 105. Their case studies of major inventions show a trend towards corporate invention and innovation similar to that in the patent statistics, though the authors are, somewhat perversely, concerned to emphasize that the individual inventor remains important (which is true) rather than that a major shift has occurred (which is striking).

[30] R. S. Sayers, 'The springs of technical progress in Britain 1919–39', *Economic Journal*, vol. 60 (1950).

tiveness of motor lorry transport, which was replacing the railway system on which prewar marketing had been largely based. The diverse Anglo-Dutch Unilever empire perhaps best exemplifies the type of integration and diversification based on the logic of marketing skills which such developments made possible.[31] This firm, the result of a series of mergers culminating in the merger of the Margarine Union and Lever Brothers in 1929, had been created primarily because of the need to coordinate the raw material purchasing policy of the constituent firms, and it held extensive interests throughout the world in oils, fats, soap and margarine markets. Yet its core of skill in marketing was also an important part of the process by which it maintained its position as an integrated food and proprietary products marketing group. In addition to consolidating the accumulated marketing experience of the constituent partners in the mergers of 1918–29 which built up the group, the company consolidated much of its road distribution network into the S.P.D. Company, developed its own advertising agency, and, through its Allied Suppliers subsidiary, controlled a large chain of retail grocery shops, including the Maypole, Home & Colonial, and Liptons groups. It was essentially by developing the skills that had been learnt in marketing soap, margarine and other traditional products that the company was later able to continue its growth by developing products such as frozen foods and a wider range of branded household goods.

Whilst no other group pursuing a strategy of marketing oriented growth could rival Unilever in size, there were many others which were developing an expertise in marketing and distribution in similar though not identical ways. Among the largest fifty firms of 1930, for example, there were Reckitts (which was to merge with Colmans in 1938), United Dairies, Boots, J. Lyons, and Cadbury-Fry. To assess the net benefits of these developing skills would be a difficult task,[32] but in so far as they achieved economies of scale in marketing and the more efficient integration of manufacturing with distribution and retailing, they represented an improvement on earlier marketing methods which had been based on a greater number of middlemen and less integrated company operations.

These categories have by no means exhausted the full range of direc-

---

[31] C. Wilson, *The History of Unilever*, vol. 2 (1954), pp. 301–99. P. Mathias, *Retailing Revolution* (1967), pp. 258–97. W. J. Reader, *Hard Roads and Highways, S.P.D. Ltd 1918–1968* (1969).

[32] E.g. there would be the problem of assessing the extent to which, by increasing barriers to entry and artificially differentiating products, their increased advertising expenditure produced welfare losses.

tions in which the corporate economy of the interwar years was expanding. This was also the period in which press barons, such as Lord Rothermere and the Berry brothers (later Lords Kemsley and Camrose), gained control of a large share of the national and local newspaper market. Whereas in 1921 the big press chains controlled only 15 per cent of newspaper output, by 1929 this had risen to 44 per cent, and the chains also integrated backwards to the manufacture of newsprint and introduced more aggressive marketing methods.[33] Whatever the social and intellectual losses these moves involved – and they were no doubt considerable – there were substantial financial gains from limiting the number of competing local newspapers, thus reducing the overlapping activities of staff, which had been a consequence of competition between many newspapers to report the same news.

Among companies ranking below the top fifty, there was also substantial growth, both internally and by merger, in the interwar years. The Metal Box Company, Tube Investments, Joseph Lucas, and Fison, for example, all owed their dominant market position to merger activity between the end of the First World War and the 1930s, a period in which they were also undergoing rapid internal growth.[34] Other groups based on marketing expertise were also being formed in this period – Beecham, for example, which was to become a leading manufacturer of proprietary articles, made important acquisitions in 1925 (Veno) and in 1938 (Eno and Macleans).[35]

Much of the technical and marketing expertise on which the growth of these large and medium-sized firms depended was generated within the firms, from their own experience of mass production and marketing and from their own research and development efforts. Yet, in view of the lagging development of Britain in some of these industries, imported expertise could also play its part and large companies were also involved in introducing new techniques in this way. The foreign influence was in some cases direct. In 1930, for example, four of the largest fifty firms – Ford, Boots, AEI, and International Nickel – were under North American ownership,[36] and smaller firms also provided a vehicle for

---

[33] Royal Commission on the Press, *Report* (Cmd. 7700, 1949), p. 193.
[34] Monopolies Commission, *Report on the Supply of Metal Containers* (1970). R. Evely and I. M. D. Little, *Concentration in British Industry* (Cambridge, 1960), pp. 249–50. Monopolies Commission, *Report on the Supply of Electrical Equipment for Mechanically Propelled Land Vehicles* (1963). Monopolies Commission, *Report on the Supply of Chemical Fertilisers* (1959).
[35] Evely and Little, *Concentration in British Industry*, pp. 189–90.
[36] See p. 120 above.

American expansion: US General Motors, for example, acquired Vauxhall in 1925 and Procter & Gamble acquired Hedley in 1930, thus gaining an important UK foothold. Whereas in 1914 there had been perhaps seventy US companies operating in the UK, by 1936 this had risen to 224 and many of these were now among the dominant firms in their industry.[37] Even where direct foreign ownership was not the avenue for transfer of technology and expertise, however, less formal means could be used to import technology from Germany and the United States. ICI, for example, had a patent sharing agreement with Du Pont and I.G. Farben (bolstered by a general agreement to divide the world market between themselves), and many of the company's ideas for diversification can be traced to a technical logic initially explored by Du Pont in the American context.[38] Among other leading firms, English Electric had a substantial licensing agreement with the American Westinghouse Corporation, Austin used American consultants when he set up his Longbridge factory for the mass production of cars, and Morris himself, when introducing the mass production of steel plates for his cars, went into partnership with an American company to finance the Pressed Steel Company.[39] It appears, then, that some of the technical progress and improved productivity of the period can be ascribed to imported technology. The large firms which were developing their expertise in these fields were playing an important part in this process, as they were also in the development of indigenous research and marketing skills.

The logic of the development of these skills was not only that large firms should import technology, however, but also that they should themselves expand overseas, to exploit their comparative advantages in wider markets. As corporations successfully attained large scale in Britain and developed a transferable expertise, therefore, they began to develop a strategy of expanding through overseas manufacturing subsidiaries.[40] The natural direction of expansion for most companies in

---

[37] J. H. Dunning, *American Investment in British Manufacturing Industry* (1958), pp. 36–57.

[38] Reader, *Imperial Chemical Industries*, vol. 2 (1975).

[39] Jones and Marriott, *Anatomy of a Merger*, pp. 140–1. PRO/BT/56/44, Chief Industrial Adviser, File 1884/7. Andrews and Brunner, *The Life of Lord Nuffield*, pp. 131–4.

[40] Research on postwar direct investment has suggested that there is a critical size below which a firm cannot envisage overseas expansion; for a contemporary business statement of the same view by Josiah Stamp, see *Macmillan Evidence*, qq. 3921, 3944–5. Cf. P. L. Payne, *British Entrepreneurship in the Nineteenth Century* (1974), p. 54.

the interwar context was the Empire. The title Imperial Chemical Industries was, for example, no accident, for, as the founders said:

> The British Empire is the greatest single economic unit in the world. . . . By linking the title of the new Company to that unit, it is intended to lay emphasis upon the fact that the promotion of Imperial trading interests will command the special consideration and thought of those who will be responsible for directing this new Company . . . and it will be the avowed intention of the new Company, without limiting its activities in foreign overseas markets, specially to extend the development and importance of the Chemical Industry throughout the Empire.[41]

Large firms like ICI, Unilever, EMI, Cadbury-Fry, Amalgamated Press, and Dunlop did develop an extensive network of overseas subsidiaries in the Empire, but neither they, nor other British firms, confined their attention to imperial areas. Unilever, for example, had extensive European and American interests, and Courtaulds, through its American Viscose Corporation subsidiary, was a dominant rayon producer in the United States, while EMI owned some fifty factories in nineteen countries, principally in Europe. While there are no reliable British estimates of the magnitude of direct investment throughout the world, we know from American studies that already by 1934 British firms had accumulated assets in US manufacturing industry valued at $305 million, and much of this was a result of expansion and new acquisitions in the 1920s.[42] No systematic study of the contribution of such overseas direct investment to the British economy in this period is available, but in 1927 the chairman of GKN gave some indication of the potential contribution to a leading company's prosperity: 'Our interests [he said] are worldwide, and I can assure you that it is only on account of that fact that we are able, in what have been twelve months of general depression, to maintain our profits on such a scale.'[43]

In contrast to the return on the mass of overseas assets accumulated by portfolio investment in the period up to 1914 – valued then at

---

[41] Reader, *Imperial Chemical Industries*, vol. 1, p. 464. See also Lord Melchett, 'Internationalism and big business', *Ashridge Journal* (Sept. 1932).

[42] US Department of Commerce, *Foreign Investments in the US* (Washington, 1937), p. 17. An unpublished study by P. D. Wright suggests that British firms made as many acquisitions abroad in the decade 1920–9 as in the whole of the period 1880–1919.

[43] *Sunday Times* (3 July 1927), quoted in G. M. Colman, *Capitalist Combines* (1927), p. 28.

£4000 million – the rate of return on the smaller sums directly invested abroad by large firms was probably greater, since it captured not only the *rentier* returns of the portfolio investor, but also the profits which accrued to the innovating entrepreneur with technical and managerial skills, of the kind now increasingly embodied in the framework of the international company. Of course, not all such ventures met with success – Morris, for example, made a disastrous attempt to enter the French motor car industry[44] – but direct investment was looked upon with increasing favour in business and official circles. The Macmillan Committee, for example, which included eminent economists, bankers and businessmen, noted that 'industry is yearly becoming more internationalized', and gave its stamp of approval to this trend, contrasting it favourably with the dominant nineteenth century forms of investment:

> ... in the realm of foreign investment it is primarily towards British-owned enterprises abroad that we should wish to see our energies and capital turned rather than merely towards subscribing to foreign government and municipal loans, which absorb our available foreign balance while doing little for our industry and commerce.[45]

The majority of the corporations whose rise we have charted, at home and overseas, were in industries which were new, or in which innovations in technology or marketing laid the base for substantial organizational changes. Yet this emphasis may be misleading,[46] for the productivity gains which lay behind the economic growth of the period were potentially as important in the traditional staple industries as in the newer sectors. Industries such as textiles, iron and steel, and shipbuilding, faced a loss of overseas markets, substantial competition in the home market, and a weak financial position, which in some cases was aggravated by overcapitalization in the postwar boom of 1919–1920.[47] It was important in these industries to tailor output to demand,

---

44 Andrews and Brunner, *Life of Lord Nuffield*, pp. 159–60.
45 (Macmillan) Committee on Finance and Industry, *Report* (Cmd. 3897, 1931), p. 165.
46 It is easier to recognize this than to rectify it, for firms in the old industries have been less active in commissioning business histories, from which information has been extensively used in the case of other industries.
47 The reduction in the number of large companies in the staple industries in the largest fifty firms of 1930 is perhaps exaggerated by this factor: their share in manufacturing output was still in 1930 somewhat higher than their share in market values on which Table 8.2 is based.

and to maximize productivity by gaining access to economies of scale, rationalization and integration; and there is some evidence that attempts of this kind were being made. Both merger activity and increases in concentration in the metal manufacture, shipbuilding and textile industries were, for example, as marked as in the new industries,[48] and substantial improvements in productivity were also achieved in these industries in the interwar period.[49]

In some cases it is possible to pinpoint particular areas in which economies flowed from the enlarged scale of firms in the staple industries, but much of the evidence currently available is impressionistic rather than conclusive. Firms like Allied Ironfounders, the Renold & Coventry Chain Company, and Jute Industries are known to have been making acquisitions and rationalizing production afterwards.[50] Cases of vertical integration were also becoming increasingly common, particularly in the iron and steel industry, and although not all firms seized these opportunities, some were able to gain economies by vertically linking the output of steel from the ore to the finished product, minimizing heat losses between the interlinked processes.[51] Stewarts & Lloyds, when building a large new integrated steel tube plant at Corby, was also able to internalize some of the external economies of its expansion there by buying up local ironstone companies, and this in itself encouraged, and increased the profitability of, such ventures.[52]

Perhaps the highest contemporary hopes for gains from rationalization were those entertained by the critics of the extremely depressed cotton trade, in which the typical unit still remained small. The logic behind these hopes was described by the Balfour Committee:

The effects of concentration and combination on the elimination of the most inefficient plant are not altogether simple. There can

---

[48] See L. Hannah, *The Political Economy of Mergers in Manufacturing Industry in Britain between the Wars* (unpublished DPhil thesis, Oxford, 1972), pp. 169–73, for merger activity; and see Appendix 2 below for concentration.
[49] Aldcroft, *The Interwar Economy*, pp. 121, 137–76.
[50] B. H. Tripp, *Grand Alliance, a Chapter of Industrial History* (1951). B. H. Tripp, *Renold Chain, a History of the Company and the Rise of the Precision Chain Industry* (1956). C. G. Renold, 'Rationalization of the management of companies under a merger', in Sixth International Congress of Scientific Management, *Development Section Papers* (1936). *Economist* (12 Jan. 1924), pp. 47–8. Board of Trade Working Party, *Report on the Jute Industry* (1948), p. 25.
[51] S. R. Dennison, 'Vertical integration in the iron and steel industry', *Economic Journal*, vol. 49 (1939).
[52] Sir Frederick Scopes, *The Development of Corby Works* (1968), p. 65.

be no doubt that the operation of free competition is a very slow and costly method for the purpose of securing such elimination. The tenacity of life shown by businesses working at a loss is sometimes extraordinary. Plant and buildings are often highly specialized, and there is a reluctance to incur an almost total loss by dismantling them while there is a chance that a favourable turn of the market will for a time at least put the business on a profitable basis. But the results of the prolonged competition of inefficient undertakings react on the more efficient, and tend to depress the whole industry; and an operation of cutting out the dead wood may be essential for the speedy restoration of prosperity and the resumption of growth from the more vigorous branches. It seems unquestionable that this operation can often be performed more speedily and 'rationally' and with less suffering through the mechanism of consolidation or agreement than by the unaided play of competition.[53]

Clearly, however, such a process of rationalization could only make a contribution to economic growth if it enabled firms to free redundant resources for transfer to other, higher productivity sectors of the economy. The Balfour Committee reflected a widespread contemporary view that this could be done more efficiently by management within a consolidated enterprise than by the alternative, traditional (but, it was felt, socially undesirable) method of long-term competitive attrition, with capitalists and workers gradually transferring their services elsewhere under the threat of bankruptcy and unemployment. The crux of this case was what might, by analogy with the more familiar economic concept of entry barriers, be thought of as 'barriers to exit'.[54] Such barriers, which induce firms to stay in an industry even though its prospects in both the long and the short term are poor, could arise from over-optimism about future prospects, or a mistaken belief that they would survive whilst their competitors would not. In the cotton industry, where these problems were met in the most extreme form, managers were able to sell output at a price below marginal cost by calling on unpaid capital from their shareholders, and even, it seems, in some cases by attracting loans from bankers. Such actions can only have

[53] (Balfour) Committee on Industry and Trade, *Final Report* (Cmd. 3282, 1929), p. 179.
[54] The following section is based on an examination of the archives of the Lancashire Cotton Corporation. For other examples of industries in which exit was hastened by mergers, see P. L. Cook and R. Cohen, *The Effects of Mergers* (1958).

diverted funds from capital formation which could have been more effectively used in the new industries, but mergers such as the Combined Egyptian Mills (fifteen firms in 1929) and the Lancashire Cotton Corporation (ninety-six firms in 1929–32) aimed instead to cut back the capital stock of these industries and release resources in the process. Despite considerable difficulties experienced in the attempt to integrate many small, competing enterprises (see pp. 84–5), the Lancashire Cotton Corporation, in particular, was able to close a number of mills, unseat the incumbent management, scrap large numbers of redundant spindles, and raise prices to marginal costs or above.

Yet the extent to which such concentration and contraction on the part of the 'old' industries enabled the 'new' industries to grow can easily be exaggerated. There remained many industries in which rationalization had made little progress, and, even without the transfer of resources from the old industries, the basic factors of production were often in ample supply elsewhere. It was an overall deficiency in demand, rather than supply constraints, that held back the further growth of the new industries. The supply of labour, for example, was not a bottleneck to growth in this period of widespread unemployment, and even if skilled workers were released by rationalization of the cotton industry, their skills were very specific and hence their contribution was unlikely to be more valuable to the new industries than that of the unskilled, unemployed workers who were available in large numbers. Capital, too, was fairly industry-specific, and whilst factories could be, and were, transferred from old to new uses, a scrapped cotton spindle or a redundant shipyard produced little but scrap metal. Thus the use in the new industries of redundant capital resources from the older trades, where it occurred, was more a convenience than a necessity. Much the same can be said about the supply of managers and entrepreneurs. While some managers undoubtedly transferred to the new industries, management still tended to be industry-specific, and the 'steel men' or 'cotton men' who transferred did not take with them more than very basic skills. The situation was, then, very different from that in the period after the Second World War when, given full employment conditions and capacity constraints on growth, the transfer of resources from the old industries was a prerequisite of the growth of the new.

Perhaps the most hopeful strategy for large firms in the staple industries was to integrate vertically or to diversify into alternative products for which demand was growing and in which their existing financial, technical and managerial skills gave them a comparative advantage. In a sense Nobel and ICI were the prime exponents of this strategy, for, from an initial base in stagnant sectors of the chemical and metals trades such as alkali, explosives and ammunition, they rationalized their production facilities in these fields and diversified into new ones, thus achieving rapid rates of growth. Yet in industries in which managerial skills were not as adaptable, perhaps for reasons inherent in technologies with little spin-off between the various branches, this strategy was not always successful. Vickers, for example, based on iron, steel and shipbuilding and, like Nobel, facing declining postwar markets, saw the need to diversify into peacetime products and in fact entered the 1920s with subsidiaries involved in electrical engineering (Metrovick) and motor vehicles (Wolseley), and also manufactured aircraft, aluminium, and scientific instruments.[55] The common dependence on metals technology does not, however, appear to have provided as useful a technical core of growth as in the Nobel case, for, despite a substantial programme of acquisition, the market value of Vickers's capital hardly increased at all between 1919 and 1930. By the later 1920s the company had abandoned much of its earlier strategy – Metrovick was sold to AEI and Wolseley to Morris, and Vickers began to concentrate more on achieving economies in its traditional spheres of operation. The result was a series of horizontal mergers with competing firms in the major heavy engineering industries. Vickers-Armstrong consolidated Vickers's shipbuilding interests with those of Armstrong-Whitworth; and the English Steel Corporation and Metropolitan-Cammell Company consolidated their steel and railway carriage interests with those of Cammell Laird. The mergers were justified by the Cammell Laird board in a letter to its shareholders, in the classic terms of the rationalization movement:

> The development of elaborate and costly machinery, the growth of mass production, the stationary if not contracting condition of the basic industries in this country, and the example of our principal competitors abroad point irresistibly to the need for far-reaching reorganization. . . . The arguments in favour of these proposals appear to us to be conclusive, for they render possible economies in production, improvements in technical efficiency and sales organization and a development of research which can be achieved in no other way.[56]

[55] J. D. Scott, *Vickers: A History* (1962), pp. 137–40.
[56] *Investors Chronicle* (22 Dec. 1928), p. 1358.

These initiatives with which Vickers was connected were by no means the only ones of a similar kind. More generally, in the steel industry, amalgamations such as Lancashire Steel, Colvilles, and British (Guest Keen & Baldwins) Iron & Steel involved the demerger of steel companies from vertical groups formed earlier, but now abandoned in a period when it was generally thought that the advantages of horizontal amalgamation were the greater.[57]

This examination of the historical record of the directions of corporate expansion in the interwar period suggests that the quantitative changes charted in the previous chapter were matched by important qualitative innovations in corporate policies, which marked a significant break with the past. It is, of course, true that many of the characteristics of the corporate economy could already be seen in the larger companies before the First World War, and the discontinuity is not by any means a total one. Kynoch, for example, had been prewar pioneers of diversification, and the United Alkali Company had had an advanced research and development programme, and both companies eventually joined ICI.[58] Firms like Lever and Bovril were developing modern techniques of advertising and marketing well before 1914, and it was on those techniques that many interwar corporations were based.[59] Some firms had also developed overseas manufacturing subsidiaries, and already in 1899 Sir Archibald Coats claimed that the greater part of J. & P. Coats's profits were derived from its overseas interests.[60] Yet, wherever quantitative assessment of trends in these developments is possible, the case for a clear shift in industrial practices in the 1920s is overwhelming. In the statistics of mergers, industrial concentration and overseas acquisitions, we have already seen that structural changes were occurring at an unprecedentedly high rate. These can be paralleled by statistics on mass production at the plant level: in 1914, for example, Ford, then the largest UK motor car manufacturer, produced only 6000 cars, whereas by 1939 six firms all had substantially higher outputs.[61] In the case of the strategy of integration and diversification also the statistics indicate a shift; between 1880 and 1918 only 8 per cent of mergers were of the diversifying or vertical kind, whereas by 1919–39 the proportion had risen to 37 per cent, and absolutely by an even

---

[57] *Macmillan Evidence*, qq. 3610, 3621.
[58] Reader, *Imperial Chemical Industries*, vol. 1, pp. 116, 147.
[59] C. Wilson, 'Economy and society in later Victorian Britain', *Economic History Review*, vol. 18 (1965).
[60] H. Macrosty, *The Trust Movement in British Industry* (1907), p. 128.
[61] Maxcy and Silberston, *The Motor Industry*, pp. 12–15.

greater amount since mergers in general were then more common.[62] Interwar corporations were also investing significantly more on research: total industrial expenditure on research and development in the later 1930s was at least £6½ million annually, compared with under £1 million before the First World War.[63] Many of the larger firms of the turn of the century merger wave had very narrow product lines and aimed primarily at the restriction of competition and the raising of prices; in the interwar period the more obvious social benefits of economies of scale and accelerated rates of technical innovation were increasingly in evidence. Imperial Chemical Industries was, in short, much nearer to the archetypal modern, diversified, multidivisional corporation than the single-product, badly organized Calico Printers Association, which might be taken as typical of the prewar stage of corporate immaturity.

Nonetheless, it could plausibly be argued in the 1930s that further opportunities for rationalization and economies of scale still existed, and many contemporary critics felt that, however far the corporate economy had advanced, it was not far enough. In some industries it could still be pointed out that British firms were still substantially smaller than their American or German counterparts. In the steel industry, for example, although by 1930 the twenty largest firms controlled 70 per cent of the British output of iron and steel, their combined output of steel was still less than one third of that of the US Steel Corporation and about the same as Vereinigte Stahlwerke of Germany.[64] Again, in the car industry, while the six largest firms accounted for 90 per cent of total output by 1939, they still manufactured forty different engine types and even more chassis and body models; and, despite opportunities for cost reduction through further standardization, they had tended to increase, rather than decrease the number of models in the 1930s.[65] Management in this, and other industries also left much to be desired: Courtaulds, for example, despite its prosperity and dominance of the rayon industry, had an inadequate managerial organization and a weak technical base.[66] The Balfour Committee bemoaned the

[62] L. Hannah, 'Mergers in British manufacturing industry, 1880–1918', Oxford Economic Papers, vol. 26 (1974), p. 11. L. Hannah, Political Economy of Mergers in Manufacturing Industry in Britain between the Wars, p. 154.
[63] Sanderson, 'Research and the firm in British industry', pp. 112, 124.
[64] Aldcroft, The Interwar Economy, p. 173.
[65] Maxcy and Silberston, The Motor Industry, p. 15. Aldcroft, The Interwar Economy, p. 184.
[66] Coleman, Courtaulds, an Economic and Social History, vol. 2, pp. 242–3.

absence of managers with a capacity to understand the technical developments which were transforming the nature of firms in the science-based industries:

> It needs to be much more generally recognized [the Committee concluded] that the requirements of industry in this respect include not only the intensive training of the expert, but also the widespread diffusion among future industrial leaders of the capacity to appreciate the value of science to industry.[67]

The economic benefits made possible by larger scale enterprise were thus neither complete nor were they to be achieved in isolation. They required complementary investment in scientific education and, more generally, a change also in social and managerial attitudes. To attempt to unravel such complementarities would be a formidable task, and this, coupled with the absence of much of the required data, makes it difficult to assess in precise quantitative terms the proportion of contemporary economic growth which can be attributed to the structural and organizational changes which were occurring. There is, however, a strong circumstantial case for believing that the structural changes and productivity improvements of the period are linked,[68] and the impressionistic evidence which we have been able to quote here tends to confirm this case. Closer studies of individual industries and firms may, in the future, provide us with more precise estimates of the gains, but for the moment we must be content with the weaker, but still significant, conclusion that the growth of large companies in the interwar years contributed substantially to the underlying productivity improvements in British manufacturing industry between the wars.

[67] (Balfour) Committee on Industry and Trade, *Final Report*, p. 218.
[68] Though in so far as concentration led to increased monopoly power and restricted output the *a priori* considerations would suggest a very different view. For a fuller consideration of this issue see Chapter 11.

9

From the 1930s to the 1950s:
continuity or change?

Private Enterprise has had the great advantage . . . of being able
to put forth a considerable range and variety of systems and
to try them out in practice. . . . it has provided us with
a fine laboratory and many experiments, the results
of which, for good and, sometimes, for evil, we
are just beginning to reap. The task . . . is to
take full advantage of what has been
going on, and to discern in
the light of these manifold
experiments which
ideas are profitable
and which
unprofitable.

LIBERAL INDUSTRIAL INQUIRY, *Britain's Industrial
Future* (1928), p. 100.

When it was suggested to Briggs that his new combination should
be called by some such title as Universally Combined General
Industries he was speechless with rage. 'We'll have the
word "Briggs" in it,' he gasped at length, 'or we'll
cancel the whole thing.'

JOHN LEE, *Letters to an Absentee
Director* (1928), p. 40.

ಙಙ

Many of the features that distinguish the modern corporate economy
from the Victorian economy of small family firms were, then, firmly
established in Britain by the early 1930s. Over large sectors of manufacturing industry the position of large integrated firms had been strengthened by vigorous internal growth and by the unprecedented merger
waves of the dozen years following the First World War. These firms
were also diversifying their product ranges through the acquisition of
businesses in related fields, and some of them had laid the basis for
continued growth by investment in new technologies and well-equipped
research laboratories. The possibilities of expansion through vertical
integration at home and abroad, and through overseas manufacturing
subsidiaries, were also being more fully explored than they had been in

earlier decades. Typically the large corporations were quoted companies and their shareholdings were widely dispersed beyond the entrepreneurial families to which most of them owed their Victorian origins. Some of them, like ICI, had begun to solve the problems of maintaining efficiency through management decentralization, and this enabled them to sustain the rapid pace of expansion, which the reduction of financial constraints on firm size and the diversification of their activities had made possible.

There are thus clear lines of continuity between the corporate economy of the 1930s and that which we now know. Moreover, the temptation to see the expansion of large corporations progressing in gradual, but decided, steps towards the high levels of concentration of today is a strong one, for the taste for rationalization and merger often seems, like a crystal dropped in a supersaturated solution, to be producing a cumulative effect of ever increasing concentration. There are a number of *a priori* considerations which lend plausibility to this view. The process by which expanding markets create new possibilities for the division of labour is a self-sustaining one, since divisions of labour also operate in their turn to expand markets, and we would expect this process, once started, to continue to open up new opportunities for economies of scale. Furthermore, given the trend towards higher market coverage, vertical integration, and diversification, the economies of such innovations would increasingly tend to be internalized by large firms. Hence external economies of scale would become progressively less important, and the repeated divisions of labour would operate more unambiguously in the direction of increasing firm size. It might also reasonably be expected that there would be some learning effect: the attractions of monopoly power, for example, would be more widely appreciated by entrepreneurs, as the experience of some firms of successfully raising profits by monopolistic consolidations became more widely known. For many reasons, then, we might expect the more rapid growth of large firms to continue. Even if no such tendency were present, however, overall concentration could still increase, owing to the operation of the 'Gibrat Effect'. If large firms only grow as fast on average as small- and medium-sized firms, there may (if there is a diversity of growth rates *within* these size groups) still be a tendency for concentration to increase, because of the operation of this effect and the absence of any compensating regressive tendencies (see pp. 6–7).

In the light of these theoretical considerations (which are also rendered plausible by the casual observer's impression of continually increasing concentration) any data which suggest that concentration has actually fallen, in any recent period, are understandably regarded with some suspicion. It was, then, natural that when, in the 1950s, British economists first attempted to measure changes in concentration over time, they reported their findings that concentration may have actually decreased in the years prior to 1950 with repeated warnings that their figures should be interpreted with caution.[1] Yet, as further evidence has accumulated, the view that the strong upward trend of industrial concentration which had been established in the 1920s did not continue in the following two decades has gained plausibility. Indeed, according to one recent estimate, the share of the top 100 firms in manufacturing output, which was as high as 26 per cent in 1930, declined to 23 per cent in 1935 and remained at that level or possibly even lower until it began to rise again in the 1950s (see Appendix 2). Another study of concentration among quoted firms in manufacturing and distribution found a similarly steep decline in concentration between 1939 and 1950, with large firms no longer sustaining their relatively rapid growth, but rather regressing towards the mean size.[2] While, in the case of both studies, there must remain doubts as to whether biases in the data exaggerate the decline in concentration, the measured decline *is* sufficiently pronounced to render it implausible that there could have been a substantial increase in concentration over these years. Hence, for manufacturing industry as a whole, the conclusion that the level of concentration in the 1930s and 1940s at the least remained stable, and may indeed have declined, is difficult to refute.

On a more disaggregated level, R. Evely and I. M. D. Little, using the *Census of Production* data for individual industries, measured changes in concentration between 1935 and 1951. They concluded that 'it is impossible to come to any definite and clear-cut conclusion about the change in the level of concentration between 1935 and 1951'.[3] This agnostic conclusion is itself suggestive of there being very little movement between these dates, and a less cautious interpretation of the evidence which they present confirms the impression that concentration, if it was increasing at all, was doing so very slowly, since the proportion

---

[1] E.g. P. E. Hart, 'Business concentration in the United Kingdom', *Journal of the Royal Statistical Society*, series A, vol. 123 (1960), pp. 51–2.

[2] P. E. Hart and S. J. Prais, 'The analysis of business concentration: a statistical approach', *Journal of the Royal Statistical Society*, series A, vol. 119 (1956). See also Appendix 2 below.

[3] R. Evely and I. M. D. Little, *Concentration in British Industry 1935–51* (Cambridge, 1960), p. 63.

of output accounted for by industries with any given degree of concentration (defined as the share of the top three firms in total employment) remained remarkably constant over the period.[4] The statistical record of merger activity in this period is also consistent with the view that dramatic changes in industrial structure were on the whole absent. In the 1930s expenditure on mergers by manufacturing firms was running at only half the level of the 1920s, and in the 1940s there was a further decline in merger activity, so that it was then more subdued than at any time since before the First World War. Not until the later 1950s was the merger intensity of the 1920s again equalled (see Appendix 1). While the various historical statistics of both mergers and concentration can, individually, pose serious problems of interpretation (which are more fully discussed in the Appendices), their collective weight again makes it difficult to evade the view that the trends established in the 1920s were interrupted for several decades thereafter. In the 1930s and 1940s the tide of the corporate economy had certainly ceased to flow; it may even (if the more speculative statistics prove to be correct) have begun to ebb.

Of course this does not imply that large firms actually contracted in size; on the contrary, they would have had to grow in absolute terms, in this period of expanding output, just in order to maintain their relative position. Equally, the general stasis did not rule out the possibility of substantial movements in individual industries: Evely and Little, for example, found that of the forty-one industries which they were able to compare directly in 1935 and 1951, there was a rise in concentration in twenty-seven and a decline in fourteen.[5] The 1930s was also a period of growth for important industrial firms such as British Plasterboard, Beecham, Bowater Paper, Hoover, Colvilles, Allied Bakeries, Morphy-Richards, and Reckitt & Colman, and the postwar years saw important consolidations such as the merger of two cable companies in 1945 to form British Insulated Callenders Cables. In the tinplate industry, also, there was the long awaited merger of the Richard Thomas, Baldwins, Guest Keen, and Llanelly interests to form the Steel Company of Wales in 1947. What was absent, however, was a vigorous and sustained merger movement leading to a substantial further concentration of output in the larger firms, such as had been experienced in 1919–30.

[4] Ibid., pp. 63–5. The problem of interpretation arises from changing census definitions.
[5] Evely and Little, *Concentration in British Industry*, p. 18.

Outside the sector of large firms, in which we have seen the origins of the modern corporate economy, however, there were substantial movements towards greater concentration of output in fewer firms. Firms employing less than 200 people, which had accounted for approximately 38 per cent of total employment and 35 per cent of net output in manufacturing industry in 1935, accounted for only 24 per cent and 20 per cent respectively by 1958;[6] and much of this decline in the small firm sector must have occurred in the 1930s and 1940s. Yet this trend is not reflected in the statistics of concentration presented earlier, for these were based on the top 100 firms in manufacturing (or the top three in individual industries), or on the quoted sector only, and thus they could not capture changes in the lower range of the size distribution of firms. What appears to have happened is that, by a combined process of competitive elimination by rivals, acquisition by other firms, and internal growth to larger size, the population of smaller firms contracted over the two decades, and births of new small firms were insufficient to redress the balance. Since the share of the largest firms in manufacturing output was also stable or declining in this period, it follows that firms in the middle ranking size ranges must have been growing in importance, and it may be inferred that it was in this sector that much of the real growth in output over the period occurred.[7] Whilst the large corporate sector showed little movement towards higher concentration in these years, then, there were, it seems, important changes making for increased consolidation and larger scale among the smaller and medium-sized ranges of firms.

Any such consolidation was, of course, welcomed by rationalizers, but, in general, contemporary observers, at least in the 1930s, regretted that the movement was not more widespread, and there were numerous

[6] (Bolton) Committee of Inquiry on Small Firms, *Report* (Cmd. 4811, 1971), p. 59.
[7] This may provide an explanation of the substantial measured decline in concentration within Hart and Prais's population of quoted firms between 1939 and 1950. Since their population (being confined to quoted companies) consists principally of the large- and medium-sized firms, a gain by the medium-sized firms relative to the large firms would appear as a decline in concentration; whereas if we were considering the whole population (as we surely should be) the matter might appear in a different light, with medium-sized firms perhaps gaining at the expense of both large and small firms. Whether we could properly account such a movement a true decline in concentration would depend on the relevant elasticity of market power with respect to firm size, the parameter α in the class of concentration measures defined in L. Hannah and J. A. Kay, *Concentration in Modern Industry: Theory, Measurement and the UK Experience* (forthcoming 1976).

calls for further consolidation among the larger firms. On the whole, however, this pressure was resisted by industrialists, and the stereotype to which this gave rise has become well known, for it became the standard critique of the conservative British businessman. He was (so ran the conventional wisdom) either ignorant of or irrationally averse to the potential benefits of rationalization, mass production and merger because he feared that they would lead to a loss of personal dominance and control. There is an abundance of examples of firms, especially family-owned firms, which might have derived financial benefits from larger scale but refused to do so for such reasons. Kenricks, the Midlands hardware firm (one of the few medium-sized companies to have been effectively scrutinized by a business historian), is a case in point. The owning family commissioned a report from their management consultants Peat, Marwick & Mitchell, which recommended that they should seek a merger with rivals. However, when the report arrived, it was rejected with the remark 'that it was a very good report but that no action should be taken', and, with friendly bankers, Kenricks were able to survive a period of low profitability without either merging, or going public.[8]

Such businesses especially disliked the prospect of being exposed to the scrutiny of the public and of financiers whom they generally (and often with justification) distrusted. As an accountant closely involved in mergers commented:

Men who have been accustomed to personal domination almost amounting to dictatorship in their own businesses, do not take kindly to a change of circumstances whereunder they merely become members of a Board of Directors, and may find themselves subject . . . to the control of others.[9]

The capitalist market ethic inherited by businessmen of this mould stressed individualist competition, rather than mutuality, in the conduct of inter-firm relations; the moral foundations of capitalist entrepreneurship implied that the individual's character and skill determined relative rewards. Their 'individualism' is hardly surprising in view of

[8] R. A. Church, *Kenricks in Hardware* (Newton Abbot, 1969), p. 213; see also pp. 216–17, 247–8, 321, 329. R. E. Wilson, *200 Precious Metal Years: A History of the Sheffield Smelting Company Limited, 1760–1960* (1960), pp. 207–208.

[9] A. E. Cutforth, *Methods of Amalgamation* (1926), p. 17. See also: J. Ryan in 'Problems of rationalization', *Economic Journal*, vol. 40 (1930), p. 365; Mr and Mrs F. H. Crittall, *Fifty Years of Work and Play* (1934), p. 162.

this ideological background. Sargant Florence well summed up its implications:

> What it actually means is, that they prefer power over their own little works to having a small share, with possibly very little power, in a large amalgamation. Secondly, they rather enjoy, as far as I can make out from talking with them and listening to them, the little game of competition; they love the secrecy of *private* enterprise and the sense of playing for a side; it is possibly a very British instinct; and they feel there is more zest in fighting against a rival than in combining with him. Thirdly, they like the feeling of running a little property; it is rather a *petit bourgeois* point of view, but after all, men in the Wool Industry or, say, the Brass Industry are *petit bourgeois* and have been since the middle of the Victorian era in England. Finally, the most important reason for their wishing to cling on to their own little business is the feudal idea of handing it on to their family.[10]

For such men – and they were represented in large areas of British industry – an increased autonomy from competitive forces in the product market, and an improved access to capital or economies of scale, both of which could become available in an enlarged quoted firm, were but poor substitutes for the greater personal autonomy and direct command relationships in their own firms.

Even in larger firms there could be a determined resistance to change based on the belief that, whatever the proven benefits in other industries, mass production and rationalization were inappropriate to their firms. Rationalizers, who complained that there was 'a national bias on the part of industry to regard its troubles as merely temporary',[11] felt that industrialists were conservatively wedded to ways that had worked in the past but were inappropriate in the present, and industrialists did sometimes act with remarkable conformity to this caricature. 'There is nothing to recommend the mass production of standardized ships,' thundered Viscount Inchcape, 'that is a method which may be described as the communism of ship construction', and he went on to advocate 'the widest imaginable diversity of detail and assembly',[12] a course which would effectively have prevented the achievement of scale economies and

[10] P. S. Florence in 'Problems of rationalization', p. 365.
[11] *Macmillan Evidence*, q. 5995; see also qq. 859, 2800, 7497. M. Webster Jenkinson, 'Memorandum on the steel trade', PRO/BT/56/2.
[12] Viscount Inchcape, 'Shipowners and shipbuilders', in *Transactions of the First World Power Conference*, vol. I (1924), p. 1115.

thus have cut right across the work of other industrialists who were attempting to introduce engineering standardization.[13] It was, in principle, open to entrepreneurs who wished to promote standardization, merger and mass production against the 'conservative' elements to acquire competitors in order to 'rationalize' their industry. However, the opportunities for entrepreneurs to act as midwives to new developments in this way were not as great as they were subsequently to become with the development of the takeover bid. In the case of private firms, of course, little could be done if the owner-directors of a firm standing in the way of rationalization did not wish to sell their assets. A sufficiently high offer could, it is true, be persuasive; and some of the more colourful press barons also used the threat of uneconomic price cutting and competitive attrition (subsidized by their other newspapers) as a means of persuading family owners of provincial newspapers to sell out their interests. Such tactics were, however, probably exaggerated in the telling, and were certainly an untypical form of takeover.[14]

In the case of publicly quoted companies, with widely dispersed shareholdings, by contrast, it might be supposed that an appeal directly to the shareholders by a bidder would be sufficient to overrule directors who refused merger negotiations which offered real economic advantages. In practice, however, takeover bids contesting the views of incumbent directors were virtually unknown before 1950.[15] The position of directors in potential bid situations was strengthened by the inadequacy of the information possessed by shareholders about the asset and profit position of their companies. This greatly weighted the advantage in favour of negotiating mergers through the directors, and even in the few cases in which direct bids were made, the shareholders invariably accepted their directors' recommendation to refuse. So remote was the possibility of a successful contested bid that it seemed quite natural for a contemporary accountant to insist that 'the negotiations must obviously be conducted by the Directors. In order to preserve proper secrecy, it is not possible for the Directors to acquaint the shareholders of the matter.'[16] This did not necessarily prevent mergers between quoted

[13] (Balfour) Committee on Industry and Trade, *Factors in Industrial and Commercial Efficiency* (1927), pp. 266–98.

[14] Lord Camrose (W. E. Berry), *British Newspapers and their Controllers* (1947), p. 17. Royal Commission on the Press, *Report* (Cmd. 7700, 1949), pp. 67–9. See also Monopolies and Restrictive Practices Commission, *Report on the Supply of Certain Industrial and Medical Gases* (1956), pp. 21–2, 92.

[15] L. Hannah, 'Takeover bids in Britain before 1950', *Business History*, vol. 16 (1974).

[16] Cutforth, *Methods of Amalgamation*, p. 37.

F

companies, for the directors (especially if they were themselves offered suitable 'compensation') might agree to a merger, and if they did so it was almost certain to go through without opposition. As the *Investors Chronicle* commentator advised: 'if the directors decide that the business should be sold it is difficult to make any other suggestion'.[17] On the other hand, where a prospective bidder could *not* convert the directors of a quoted company to his view, he was for practical purposes no more able to enforce his view in a takeover bid than he would have been in the case of an entirely private family-owned company. Indeed, the point may be made more strongly, for, while a family company might accept a generous offer because it was financially profitable to do so, directors could and did refuse offers which their shareholders would have been glad to accept. Hence, in this sense, it was more rather than less difficult to induce a quoted company to merge, and this is reflected in the merger statistics: less than 5 per cent of the firms quoted on the London stock exchange in 1924, for example, were acquired by other quoted firms in the following fifteen years, a much lower percentage than that which has prevailed since the development of takeover bidding.[18]

An entrepreneur who was convinced of the potential benefits of rationalization and mass production was, then, in a difficult position if he could not persuade other industrialists to agree to his proposals. The one option that remained open to him was to build his own plant on a large scale, and, by competitive price reductions (made possible by the use of superior, low-cost plant), to drive out his competitors (or force them eventually to accept a merger). This is precisely what happened in some industries. William Morris, for example, acquired a number of small competitors and suppliers, but owed his successful introduction of mass production not to this but to his ability, through cheaper prices, to win new markets and force small-scale specialist motor manufacturers out of business. More generally the disappearance of many small firms over manufacturing industry as a whole is in part the result of such a process of natural selection through competition. Yet such a strategy for introducing mass production was not always possible, for it depended on the existence of special conditions. In the car industry, for example, rapidly expanding markets meant that the majority of consumers were 'new' and had no initial loyalty to any particular make of car; it was, moreover, a highly competitive market. Hence Morris and

---

[17] Quoted in Hannah, 'Takeover bids in Britain before 1950', p. 72.
[18] Hart and Prais, 'The analysis of business concentration: a statistical approach', p. 169.

the other major mass producers of cars could be confident of finding customers for their output. In many other markets, however, there were imperfections of competition, resulting variously from branding, advertising, consumer loyalty, restrictive agreements or, in the case of some products, from vertical integration between a manufacturer and his suppliers or retailers. The classic case of the latter type of market imperfection was in brewing, where the tied house system and legal restrictions on the licensing of new pubs tied the brewer to a limited number of pub outlets. Any conceivable price reductions made possible by scale economies in new plant would not, in this industry, be able to capture more than a small proportion of the trade of the publicans already tied to other breweries. In this and other industries facing imperfect markets, therefore, an entrepreneur introducing mass production could not, unless he made economies large enough to permit very substantial price reductions, hope immediately to win a sufficiently large market to justify his investment. A precondition of the building of a large plant in such a market would, then, be a merger of interests with existing manufacturers, amalgamating, as it were, not only their assets but also their consumer demand.[19] In these conditions, then, the absence of a takeover mechanism might seriously inhibit desirable structural changes.

In the light of these factors, the view that British businessmen were in general forgoing the potential social benefits of reduced industrial costs through mass production (and, for that matter, the potential private benefits of monopoly profits) becomes more intelligible, for the economic mechanisms which are conventionally thought of as penalizing such ignorance or irrationality operated only perfunctorily in the interwar context. There are, however, alternative, quite plausible explanations of the apparent aversion of large numbers of entrepreneurs to larger scale enterprise. We must, then, pause before accepting the condemnations of the rationalizers – who frequently had little knowledge of the conditions faced by the industries they criticized – to ask whether the actions of industrialists can rather be explained, in part at least, as a rational response to the economic conditions of the time.[20]

[19] A similar remark was made by the Chairman of Stewarts & Lloyds, that 'concentration of production could not take place without concentration of demand' (Annual General Meeting, 18 May 1938). Cf. F. M. Scherer, *Industrial Market Structure and Economic Performance* (Chicago, 1970), pp. 116–18.

[20] A parallel case is the widespread criticism of British businessmen of the late nineteenth century for not adopting 'modern' techniques (ring spindles,

It is, for example, worthwhile recalling that the 1930s and 1940s were an exceptional period in economic history, encompassing the most serious world depression and the most destructive world war ever experienced. It is conceivable that such upheavals interrupted the long-run trend towards increasing competitive advantages for larger scale enterprises, and that this, rather than entrepreneurial irrationality, explains why the corporate economy changed down to a lower gear. Closer consideration suggests that this might indeed be the case.

If world economic conditions rather than an outbreak of irrationality had temporarily halted the trend towards higher concentration, for example, the effect would have been widespread and other industrial nations would have experienced declining or static concentration: the evidence for the United States confirms that such a trend was also found there.[21] The expectation of a downward movement in concentration is also supported by a consideration of the likely impact of a world depression on the determinants of concentration which have been identified. In markets in which demand had contracted, for example (such as those faced by the British export industries in this period), the scope for economies of scale may have been attenuated, given the classical proposition that the division of labour is limited by the extent of the market. Firms that saw in their market environment no prospect of a sufficiently large demand to enable them to reap the scale economies of low-cost mass production could then sensibly reject, on marketing grounds, mergers which on technical, engineering grounds would have been appropriate.[22] Financial economies of scale, which had also been acting to increase the size of firms, were also reduced as the collapse of international investment in the 1930s led City institutions to turn their

Bessemer steelmaking, etc.) as quickly as foreign competitors. Interestingly, subsequent analysis by the 'new' economic historians has shown that in some cases such 'failings' were a rational response to relative factor prices faced by businessmen in Britain at the time; see e.g. D. N. McCloskey and L. Sandberg, 'From damnation to redemption: judgements on the Late Victorian entrepreneur', *Explorations in Economic History*, vol. 9 (1971). See also N. K. Buxton, 'Entrepreneurial efficiency in the British coal industry between the wars', *Economic History Review*, vol. 23 (1970), pp. 476–97, and the discussion in *Economic History Review*, vol. 25 (1972), pp. 655–73.

[21] M. A. Adelman, 'The measurement of industrial concentration', *Review of Economics and Statistics*, vol. 33 (1951).

[22] E.g. PRO/BT/55/49; P. W. S. Andrews and E. Brunner, *Capital Development in Steel* (Oxford, 1951), p. 172. The suspicion remains, however, that in many such cases the anticipated market was not sufficiently large because of imperfections in the market and an unwillingness to overcome them through merger, rather than because of an overall deficiency in demand.

attention to the financing of British industry.[23] Thus William Piercy, who had in the 1920s advised a company to seek outside capital only at the £250,000 level, could in the 1930s advise them to start at £100,000, or even, on the provincial exchanges, at £40,000; and later, in 1945, the foundation of the Industrial and Commercial Finance Corporation, explicitly to cater for the smaller companies, further improved their position.[24]

Of course, depression had not limited markets and reduced the potential economies of scale everywhere, and there were many markets in the 1930s in which demand was growing, but, even here, economies of scale were not leading inexorably to large-scale enterprise. Some economies of scale could, for example, be achieved by the collective action of independent firms rather than by further consolidation. Industrial research associations, jointly financed by industry and government, were in this period allowing smaller firms to gain the advantages of research which, acting independently, they would have been unable to afford.[25] In marketing, also, there remained advantages for the small, specialized unit which was quick to respond to changes in consumer tastes, and in the production technology of many industries economies of scale were unimportant or had been exhausted at quite a small size.[26] Hence it may well have been that many industries had, by 1930, achieved a position in which many of the major economies of scale related to firm size were exhausted.

One of the factors that continued to limit overall economies of scale was the widespread experience of management difficulties in coping with large amalgamations, and this continued to militate against rapid company growth. Firms tried in a number of ways to avoid these difficulties, for example by eschewing multi-firm amalgamations and developing new organization structures (see Chapter 6), but such managerial initiatives were not universal amongst large firms, and those

---

[23] A. T. K. Grant, *A Study of the Postwar Capital Market* (1937), pp. 189–96.

[24] W. Piercy, 'The financing of small business', *British Management Review*, vol. 3 (1938). R. Frost, 'The Macmillan gap 1931–1953', *Oxford Economic Papers*, vol. 6 (1954).

[25] H. F. Heath and A. H. Hetherington, *Industrial Research and Development in the United Kingdom* (1946).

[26] I. F. Grant, 'The small unit in industry', *Economic Journal*, vol. 32 (1922). Anon., 'Why we did not join the amalgamation', *Business* (Jan. 1928), p. 15. J. Jewkes, 'A statistical study of the economies of large-scale production', *Transactions of the Manchester Statistical Society* (1931–2). But cf. L. Rostas, *Productivity Prices and Distribution in Selected British Industries* (Cambridge, 1948), pp. 28–30, for evidence of increasing returns over a wide field of industry in 1935.

which did not devise appropriate structures could encounter very serious managerial problems. Some large companies found that they had expanded more rapidly than could conveniently be handled by their existing managerial organization, and demergers to cope with this were sometimes necessary.[27] In more serious cases bankruptcy or financial reconstruction was forced on firms which suffered from overcapitalization and an over-rapid rate of expansion. Evidence of such cases is not readily available, for failing firms leave few records, but one recent study,[28] which examined seven of the larger mergers of 1919–28 (each of them involving five or more firms disappearing, valued at over £3 million), found a high incidence of failure, and a significant difference between the successes and the failures was the ability of the successful firms to create a viable management structure. In view of these experiences, and more general managerial disquiet at the problems of rapid expansion, it is hardly surprising that there were in the 1930s only two mergers of comparable magnitude: the acquisition by Wallpaper Manufacturers of five of its competitors in 1934, and the consolidation of six sugar beet processing firms to form the British Sugar Corporation in 1936. Significantly these had the additional advantages of, respectively, a well-entrenched monopoly position and a large government subsidy, and this no doubt in part accounts for the willingness of the industrialists involved to embark upon them.[29] In other cases, however, businessmen were increasingly wary of new, poorly planned merger schemes, and those companies which had been able to expand rapidly in the 1920s often had their hands full with the problems of establishing managerial controls over their already enlarged enterprises. Hence for most large companies the managerial constraint on their growth remained a real one.

This is not, of course, to say that the advantages of merging were no longer sought after; on the contrary, the perennial attractions of monopoly power were still much appreciated by industrialists. There was, however, a shift in emphasis in the 1930s, a shift which can be seen both in the rationalization literature and in business policy. The shift may be

[27] C. Wilson, *The History of Unilever*, vol. 1 (1954), p. 298. J. D. Scott, *Vickers: A History* (1962), pp. 167–8. A study of large companies between 1919 and 1939 showed that for every seventeen subsidiaries which they acquired, one subsidiary was demerged; see L. Hannah, *The Political Economy of Mergers in British Manufacturing Industry between the Wars* (unpublished DPhil thesis, Oxford, 1972), p. 270.

[28] L. Hannah 'Managerial innovation and the rise of the large-scale company in interwar Britain', *Economic History Review*, vol. 27 (1974), pp. 267–9.

[29] Ibid., p. 269.

broadly characterized as one away from merger and the growth of large firms towards the restriction of the market by collective agreements.[30] This option had the added attraction that it bypassed the managerial problems of enlarged companies since it maintained the independent status and the existing size of firms intact, but it could, through an efficiently organized cartel, gain the monopolistic advantages of market control, allowing firms to restrict output and increase prices above competitive levels. Of course, this option had been open to businessmen for a long time and had been extensively used in the past, but a number of factors increased its attractiveness and viability in the 1930s. The common law, which had for some time tolerated restrictive agreements, showed increasing favour to such agreements in the interwar period: 'a definite movement from the protection of individual economic freedom to the recognition of the legitimate purposes of group control'.[31] For reasons very similar to those which had gained the rationalization movement support for larger scale enterprise, collective action by producers through trade associations was increasingly represented as being in the national interest. Rationalizers often stressed the virtues of 'organized marketing' and there were pressures on the government to give more support to private restrictive agreements, including the portentously named proposals for the 'self-government of industry' canvassed by Political and Economic Planning and The Industrial Reorganization League.[32] The latter was backed by Lord Melchett of ICI, but industrialists could not agree amongst themselves that the general enabling bill which he proposed was desirable, and the Cabinet in fact rejected the proposals (which had the support of Conservative backbenchers) because it 'would have the effect of checking industrial enterprise and development'.[33] Nevertheless, the National government did

---

[30] See, generally: A. F. Lucas, *Industrial Reconstruction and the Control of Competition* (1937); Monopolies and Restrictive Practices Commission, *Collective Discrimination* (Cmd. 9504, 1955); Political and Economic Planning, *Industrial Trade Associations* (1957).

[31] W. Friedmann, 'The Harris tweed case and freedom of trade', *Modern Law Review*, vol. 6 (1942), p. 13. A. L. Haslam, *The Law Relating to Trade Combinations* (1931).

[32] Lucas, *Industrial Reconstruction and the Control of Competition*, pp. 55–6. H. Macmillan, *Winds of Change 1914–1939* (1966), pp. 370–2. A. Marwick, 'Middle opinion in the thirties, planning, progress and political agreement', *English Historical Review*, vol. 79 (1964).

[33] PRO/CAB/17/35 (discussion of Mr Molson's motion). Similar proposals in the postwar period were also rejected, this time by industrialists fearing the Labour government's intervention in their affairs through such state-backed cartels.

give some support to more limited restrictionist schemes. The 1935 budget granted a tax concession to voluntary schemes designed to restrict capacity.[34] This encouraged many private schemes on the model established by the National Shipbuilders Security scheme, which had earlier been encouraged by the Bankers Industrial Development Company in order to rationalize capacity in that industry.[35] Also, in the cotton industry, where there had been a decade of largely unsuccessful private experiments with schemes for cutting down capacity and reducing competition, the government gave legislative sanction to a levy for the scrapping of spindles.[36] The Import Duties Advisory Committee also fostered schemes for the control of investment and pricing in steel and elsewhere, sometimes using the threat of withdrawal of the tariff as a means of enforcing its views on the industry.[37] This supportive policy reached its zenith in the war, when the government gave more solid sanction to trade associations and restrictive agreements as part of wartime output planning.[38] Whereas in the First World War wartime pressures had led to increasing concentration, the system of physical controls, raw material rationing, and output planning in the Second World War, by contrast, was designed 'to alter the basic structure as little as possible'.[39]

Even where positive government initiatives were lacking, the private associations and industrial cartels which were formed in the 1930s and during the war appear to have been both more permanent and more successful than many of their predecessors, which had often foundered in trade depressions, as secret price shading developed and dissension between members mounted. Partly they owed their new-found prosperity to the adoption of a general tariff in 1932, which reduced the threat of import competition from abroad that had previously inhibited

[34] Finance Act, 1935, s. 25.
[35] Lucas, *Industrial Reconstruction*, pp. 55–6.
[36] The Cotton Industry (Reorganisation) Act, 1936. A further act in 1939 envisaged price restriction in the cotton industry but was never fully implemented.
[37] Sir Herbert Hutchinson, *Tariff Making and Industrial Reconstruction* (Oxford, 1965).
[38] D. H. Aldcroft, 'Government control and the origin of restrictive trade practices in Britain', *Accountants Magazine*, vol. 66 (1962).
[39] P. J. D. Wiles, 'Pre-war and wartime controls', in G. D. N. Worswick and P. H. Ady (eds), *The British Economy 1945–1950* (Oxford, 1952), p. 151. See also: G. C. Allen, 'The concentration of production policy', in D. N. Chester (ed.), *Lessons of the British War Economy* (Cambridge, 1951), pp. 167–81; Evely and Little, *Concentration in British Industry*, pp. 178–180.

monopolistic pricing[40] and thus greatly strengthened the position of cartels in the home market. Partly also they were strengthened in the 1940s because wartime and postwar shortages created a sellers' market so that there was little competitive pressure to act as a disruptive force on the cartels. But in addition to these factors it is surely no accident that the new success of restrictive agreements followed on the rapid increase in concentration which industry had experienced in the 1920s. It is a commonplace that the difficulties of organizing a cartel increase in proportion to the number of its members: if there are many firms in an industry the costs of policing an agreement are high, and there are dangers that 'free riders' will enjoy the benefit of high prices without reducing their production proportionately. With more mergers between firms and increased concentration, however, both formalized cartels and informal oligopolistic collusion were substantially simplified and the risks of failure were reduced.[41] In other respects, also, the large corporate firms of the 1930s were better placed than their predecessors to reap the advantages of oligopoly situations. Because of improved techniques of market research, for example, they were better able to exercise price discrimination or to raise prices to a profit-maximizing level, without a further movement to a full monopoly position, or even, perhaps, without a formal price agreement.[42]

The significance of this increasing reliance of British companies on market control through formal cartels or informal oligopolistic collusion, rather than through larger scale companies, can be seen in the contrast between industrial attitudes in Britain and in the United States, where, despite increasing official tolerance during the 'New Deal', cartels still remained unpopular and were conducted with difficulty and often in secret. The contrasting national attitudes were revealed in the negotiations in Britain of the American industrialist Gerard Swope, of the International General Electric Company. When Swope (who already controlled AEI) approached the other major British electrical manufacturers with a view to acquiring control of their undertakings, his financial advisers found it difficult to see why he should prefer this to the alternative of a monopolistic cartel. Vivian Smith, his English

---

[40] Cf. (Balfour) Committee, *Factors in Industrial and Commercial Efficiency*, p. 70.
[41] Lucas, *Industrial Reconstruction and the Control of Competition*, p. 176. G. J. Stigler, 'A theory of oligopoly', *Journal of Political Economy*, vol. 72 (1964).
[42] L. F. Urwick, *The Meaning of Rationalization* (1929), pp. 39, 86–9. L. Barley, *The Riddle of Rationalization* (1932), ch. 9.

adviser, put the matter clearly in a note to the Lazards banking house (which was advising one of the companies on Swope's 'shopping list'): 'Personally [he wrote], I am not at all sure whether instead of having one great octopus in the trade it is not better to have several big companies with a close working arrangement.'[43]

Such an arrangement would, of course, have all of the monopolistic advantages of a merger, yet it would avoid the organizational problems which would be created for the management of a 'great octopus'. In the event, the other major English companies – GEC, AEI, and English Electric – appear to have agreed with Smith: certainly the industry was organized on this basis for almost three decades, before the cartel finally collapsed and was eventually replaced by a merger in 1967–8.

It is because of the great expansion in the number of successful cartels and restrictive agreements in the 1930s and 1940s that that period has been commonly seen as the highwater mark of the restrictionist and anti-competitive forces in British industry. On the whole this view is a just one,[44] but the stronger version of it, which goes on to ascribe many of the ills of inefficiency in British industry to the euthanasia of competition, is more problematical, for, despite much contemporary evidence of inefficiency, it is by no means obvious that this was uniquely concentrated in the monopolistically organized industries.[45] Moreover, it is a moot point whether the extreme competitive struggle which, in the absence of cartels, would have been forced on industries suffering from overcapacity would have served these industries better than collective schemes for capacity reduction. In cotton spinning, for example, the financially weak firms were often those which had suffered from speculative company promotions and not necessarily those with uneconomic equipment; in this industry, therefore, the competitive process working through bankruptcy might have had adverse effects on overall productive efficiency, had it not been tempered by planned scrapping of the uneconomic equipment.[46] Equally the pessimistic view of the period is not

---

[43] R. Jones and O. Marriott, *Anatomy of a Merger* (1970), p. 124. For similar views on other industries, see: Church, *Kenricks in Hardware*, p. 161; F. Lee, 'Will the big business last ?', *Ashridge Journal* (Sept. 1933).

[44] Though in some industries problems of overcapacity actually intensified competition and there were other longer term developments working for increased competition, cf. pp. 188–90 below.

[45] There appears to be no clear relationship between productivity performance and levels of concentration in industries during this period; see, e.g., C. F. Carter and B. R. Williams, *Industry and Technical Progress* (1957), p. 121.

[46] Cf. G. C. Allen, 'An aspect of industrial reorganisation', *Economic Journal*, vol. 55 (1945).

entirely borne out by the evidence of further improvements in the management of large companies, the rationalization of plants, and the move towards standardized mass production methods in British firms of the 1930s. English Electric, for example, which had languished between its formation in 1919 and 1930, was revitalized in the 1930s under new management.[47] Other large corporations were also reaping the advantages of their earlier investments in research: in 1933, for example, ICI's research chemists discovered polythene, and EMI was at the same time developing television technology, which led in 1936 to the setting up in Britain of the first electronic television system in the world.[48] The trend towards mass production on the plant level also shows clearly in the statistics for these years. In manufacturing industry as a whole, the proportion of total employment in the largest plants (those employing 1500 or more workers) rose from 15 per cent in 1935 to 24 per cent in 1951, and the number of smaller plants (those with 200 or fewer employees) declined proportionately: the percentage of employment in these plants fell from 44 per cent in 1935 to 35 per cent in 1951.[49] This trend to larger plants was particularly noticeable in the chemicals, metal manufacture, building materials, aircraft and engineering industries, where literary evidence confirms that there were substantial economies of scale. Even without a further concentration of output in fewer *firms* then, there was a clear trend towards the concentration of output in fewer *plants;* and, from the point of view of production economies of scale, it is these latter changes, rather than the earlier changes in the structure of firms, that have the greater significance. Moreover, the two phenomena are surely not unrelated. When large firms were first created by a merger the management would be able to reap some of the advantages within a very short time. The cost of specialized central services could be spread over all of the subsidiaries, competitive price cutting by different branches of the firm could be prevented, and hence perhaps market prices could be raised. It would sometimes also be possible to 'rationalize' sales forces, product ranges, advertising budgets, and the internal financial and accounting functions. Sometimes small plants could be closed down and production concentrated in the efficient

---

[47] Jones and Marriott, *Anatomy of a Merger*, pp. 130, 138. 'New management raises profits £200,000', *Business* (Apr. 1932), pp. 7–8.

[48] J. Jewkes, D. Sawers and R. Stillerman, *The Sources of Invention* (Oxford, 1958), pp. 339–42, 387–8.

[49] A. Armstrong and A. Silberston, 'Size of plant, size of enterprise, and concentration in British manufacturing industry 1935–58', *Journal of the Royal Statistical Society*, series A, vol. 28 (1965). (Bolton) Committee, *Report*, p. 59.

(and usually larger) plants, which had previously been working below capacity. Other scale economies, however, could only gradually be reaped, since they were not matters of rearrangement of existing assets and working practices, but had rather to be embodied in new investment on a larger scale. Of course, it might be argued, from an engineering point of view, that a merger cannot make any difference to such an investment decision, since all firms, large or small, face the same long-run production function, and will thus choose to build their new plants of the optimal size, irrespective of their own initial sizes. However, there are a number of reasons, implicit in our earlier analysis, which would lead us to expect a merged firm which had successfully attained larger scale (such as those formed in 1919–30) to build larger plants than would have been planned in the investment programmes of the constituent firms, had they been acting separately. Given widespread competitive imperfections in the product market, for example, a single firm which built a large new plant in order to gain access to economies of scale could not be sure of an appropriately enlarged share of the market. Hence in some industries, where economies of scale existed, but the resulting price reductions were not sufficient to enable a firm to wrest established market shares from competitors, a merger was a precondition of larger plants (see p. 151, above). Moreover, given imperfections in the capital market, and the extensive reliance on the ploughback of profits for the finance of investment, a larger merged company would also be better able to finance the capital investment required for large plants.[50] We may, then, expect the consequences of a substantial increase in the number of larger firms (such as had occurred in the 1920s) to be seen not only in immediate efficiency gains (and monopolistic exactions) but also in the longer run, as such larger plants, bringing with them new economies of scale, are being built. Some indication of the delay between the initial merger wave and the reaping of such efficiency gains can be seen in the statistics of the proportion of manufacturing net output produced in small plants (defined as those with less than 200 employees). Between 1924 and 1935 (when the initial direct effects of the 'rise of the corporate economy' were presumably being felt) the share of output produced in such plants was only reduced from 42 to 41 per cent. In the following sixteen years, however, this share was substantially reduced, so that by 1951 small plants accounted for only 32 per cent of manufac-

[50] See pp. 71–6 above. Mergers may also facilitate investments on a larger scale by reducing uncertainty in investment planning; see G. B. Richardson, *Information and Investment* (Oxford, 1960).

turing output;[51] and much of this can be ascribed to the building of larger plants by the larger firms created in the merger wave of the 1920s.[52] The rapid recorded increase in industrial productivity in the 1930s and the resilience of productivity in the 1940s may, then (in so far as they were due to plant economies of scale), be an important, but delayed, benefit of the earlier industrial movements which had led to the rise of a corporate economy of firms capable of making larger scale investments.[53]

A favourable account of the industrial developments of the 1930s and 1940s might, then, be constructed from the evidence of structural change in these years. It would be admitted that large firms were no longer gaining relative to others, but this could be represented as a rational entrepreneurial response to the managerial difficulties of large-scale enterprise and to the exhaustion of the economies of scale available to them at the level of the firm. At the level of plants, however, the benefits of the earlier consolidations were being pursued and larger scale plants were producing continuing productivity gains. Yet such a favourable view would be unlikely to be universally accepted, and a very different picture, with a pessimistic viewpoint, can also be drawn. On this view, firms were not pursuing the rationalization of capacity and the introduction of mass production methods as vigorously as they might properly have done, because of the absence of a competitive spur. Between 1935 and 1951, for example, industrialists commonly met an increase in demand by increasing the total number of plants in operation (rather than, as later, by building new plants on a sufficiently large scale to permit a net decrease in the number of plants), thus presumably forgoing greater economies of scale.[54] Instead of seeking profits in mass

---

[51] (Bolton) Committee, *Report*, p. 59.

[52] Armstrong and Silberston, 'Size of plant, size of enterprise, and concentration in British manufacturing industry 1935–58', pp. 398, 408, suggest that most of the corresponding increase in the share of output came from plants with 1000 or more employees. Of course not all of these were built by the largest firms: some would have grown from a smaller size and still be owned by relatively small firms. See also Evely and Little, *Concentration in British Industry*, pp. 165–75, 181–3.

[53] K. S. Lomax, 'Growth and productivity in the United Kingdom', *Productivity Measurement Review*, vol. 38. W. E. G. Salter, *Productivity and Technical Change* (Cambridge, 1966).

[54] Between 1935 and 1951 the increase in output was met by an increase of 18 per cent in the number of plants; but between 1951 and 1958, when output again rose, the number of plants actually declined; see Armstrong and Silberston, 'Size of plant, size of enterprise, and concentration in British manufacturing industry, 1935–58', p. 408. Of course, it could be argued that

production economies, the pessimistic argument might run, they increasingly sought higher returns through monopolistic restrictions, and the cartels they formed could only secure the private advantages of monopoly power, rather than those benefits which also corresponded to a social advantage, such as economies of scale in production. The prevalence of restrictive practices, and of large firms which were little more than loose confederations of subsidiaries, lends weight to the view that it was a desire for monopoly, rather than for real economies, that was behind the industrial developments of these years.[55]

That we have found evidence for both of these views can hardly be surprising, for it is unlikely that such generalizations can be anything other than oversimplifications of the diverse experience of many entrepreneurs, with varied levels of skill and working in a range of industries, each with different economic conditions. The game can be played of producing examples of poor performers and of highly successful ones, but an objective assessment is only possible if adequate yardsticks for performance can be devised and if full evidence is available on the conditions faced in each industry. In the absence of such studies, we must for the present rest content with the limited conclusion that, while there are intelligible reasons for expecting poor performance in this period – monopolistic imperfections and the absence of the take-over bid mechanism for disciplining quoted companies, for example – there is also some indication that some of the longer-run benefits of the earlier merger movement were also being exploited as new investment in larger plants embodied improvements of productivity. For the more restless contemporary critics, however, such a limited and fence-sitting conclusion was unconvincing, and in the later 1940s the view was again gaining ground that the available economies of scale had not been sufficiently exploited. Again American methods seemed to point to changes in British technique. It was this view that stimulated Sir Stafford Cripps to establish, in August 1948, the Anglo-American Productivity Council. Selected teams of businessmen and trade unionists were sent to view US industry under its auspices, and they unanimously stressed the need to emulate American methods. At the same time academic research was confirming that US industry achieved consider-

---

there were just fewer scale economies available in the earlier period. Cf. J. Jewkes, 'The size of the factory', *Economic Journal*, vol. 62 (1952).

[55] A number of Monopolies Commission reports ascribe the failure to rationalize in these years to monopolistic practices and the absence of competitive pressure. On firms which were loose confederations of subsidiaries, see pp. 97–8 above.

ably higher levels of production per head than UK industry, and it was noted that the US superiority was most marked in the mass production industries. Whilst differences in size of plants as such could not in general account for the gap in performance, US firms were found to be more conscious of the need for long runs, standardization and expenditure on research, and the businessmen on the postwar productivity missions were in general impressed by American achievements, as their predessors in the rationalization movement of the 1920s had been.[56] If critical self-analysis is a virtue, then virtue was once again widespread in the later 1940s, as business criticism of Britain's industrial structure mounted. As in the 1920s, this reassessment was again to have profound implications for industrial practices and the structure of firms.

[56] N. Leyland, 'Productivity', in G. D. N. Worswick and P. H. Ady (eds), *The British Economy 1945–1950* (Oxford, 1952), pp. 381–98. L. Rostas, *Comparative Productivity in British and American Industry* (Cambridge, 1948), pp. 60–63.

IO

The modern corporate economy

... There was a need for more concentration and rationalization
to promote the greater efficiency and international
competitiveness of British industry. The
changes which had so far taken place
in this direction ... did not yet
match the economy's
requirements.

INDUSTRIAL REORGANIZATION CORPORATION, *First Report
and Accounts* (1968), p. 5.

Size in itself is no solution – indeed it is not
without its disadvantages. ...

INDUSTRIAL REORGANIZATION CORPORATION, *Report
and Accounts for the Year ending
31 March 1969* (1969), p. 7.

ఇచ్ఛ

From the later 1940s the interest of government policy makers in the
subject of industrial organization quickened, and the significance of
past changes in the structure of firms (and the possibility that the trends
established earlier would become even more marked) increasingly
attracted the attention of economists and businessmen also. From this
time on, the growth of large firms and the rise in merger activity is
paralleled by an equally rapid rise in the volume of literature devoted
to the subject. The impact of this literature has been wide. It has led
to a re-evaluation of the Labour Party's approach to business and a new
interest in planning through large-scale enterprise.[1] On the theoretical
level, traditional models of economic behaviour have been questioned,
and attempts have been made to modify them to take account of the
development of 'managerial' capitalism.[2] There has also been a great
deal of empirical work on the dimensions of merger activity, industrial
concentration and business behaviour.[3] Yet contemporary writers,

[1] C. A. R. Crosland, *The Future of Socialism* (1956). A. Shonfield, *Modern
Capitalism* (1965).
[2] R. Marris, *The Economic Theory of 'Managerial' Capitalism* (1964).
[3] D. F. Channon, *The Strategy and Structure of British Enterprise* (1973).
K. D. George, *Industrial Organisation: Competition, Growth and Structural*

understandably impressed by the sheer weight of information produced
on postwar mergers and increasing concentration, have sometimes
concluded from it that modern developments were historically unprecedented, a view which is, as we have seen, erroneous. It may, then,
be worthwhile to review the evidence for recent years in the light of our
analysis of similar phenomena in earlier periods.

There can be no doubt that, after the pause in the 1930s and 1940s,
concentration again increased at a rapid pace in the 1950s and 1960s.
The share of the largest 100 companies in manufacturing net output,
which in 1953 (as in 1930) stood at 26 per cent, had by 1968 risen to
42 per cent, and this movement appears still to be continuing.[4] In
individual industries, also, *Census of Production* data confirm that the
dominant tendency was for the largest firms to increase their market
shares in all three of the inter-censal periods 1951–8, 1958–63 and
1963–8.[5] As in the earlier period of rapidly rising concentration, this
more recent movement was associated with intensive waves of merger
activity affecting a wide range of industries. Annual expenditure by
manufacturing firms on mergers (which had been relatively low for
almost three decades) had regained the high levels of the 1920s (in real
terms) by 1956, and reached new heights in the merger boom which
peaked in 1968 (see Appendix 1). The significance of this merger wave
in consolidating the population of firms can be measured in the listings
of the larger (principally quoted) companies in UK manufacturing.
The listings, compiled by the Board of Trade and its successors, include
firms accounting for about two-thirds of manufacturing output.[6]

Change in Britain (1971). P. E. Hart, M. A. Utton and G. Walshe, *Mergers
and Concentration in British Industry* (Cambridge, 1973). G. D. Newbould,
Management and Merger Activity (Liverpool, 1970). A. Singh and G. Whittington, *Growth Profitability and Valuation: A Study of United Kingdom
Quoted Companies* (Cambridge, 1968). A. Singh, *Takeovers: Their Relevance
to the Stock Market and the Theory of the Firm* (Cambridge, 1971). M. A.
Utton, 'Mergers and the growth of large firms', *Bulletin of the Oxford University Institute of Statistics*, vol. 34 (1972).

[4] Appendix 2. The increase in absolute terms of 1·1 percentage points per
annum between 1953 and 1968 may be compared with the increase of 0·8
percentage point per annum between 1919 and 1930. In relative terms, however, the rate of increase was very similar in both periods; cf. p. 105 above.

[5] K. D. George, 'Changes in British industrial concentration 1951–58', *Journal
of Industrial Economics*, vol. 15 (1967). M. C. Sawyer, 'Concentration in
British manufacturing industry', *Oxford Economic Papers*, vol. 23 (1971). Provisional information on the 1968 Census from the Business Statistics Office.

[6] The population is the same as that used in L. Hannah and J. A. Kay, *Concentration in Modern Industry: Theory, Measurement and the UK Experience*
(forthcoming 1976).

TABLE 10.1   *The increasing concentration of net assets in UK manufacturing industry, 1957–69*

	1957	1969
Number of firms	1182	744
Share of the largest 5 firms	17·0%	20·1%
Share of the largest 50 firms	48·4%	60·9%
Share of the largest 100 firms	60·1%	74·9%
Share of the largest 200 firms	73·0%	86·2%

Source: 1957: Board of Trade, *Company Assets and Income in 1957* (1960). 1969: Unpublished Department of Trade and Industry listing.

Using net assets as a measure of company size, various concentration ratios can be calculated for this population in both 1957 and 1969 and these are shown in Table 10.1. While these increases in concentration are less dramatic than those within the population of larger firms in the 1920s (see p. 112), they confirm the impression of a substantial further concentration of manufacturing assets into fewer and larger industrial groupings. Moreover, over the period as a whole, mergers were the major cause of increasing concentration, as they were in the 1920s. Indeed one study has suggested that concentration might actually have declined in this period if it had not been for the concentration-increasing effects of the merger waves.[7] Parallel to this movement towards larger enterprises, small firms continued their almost uninterrupted decline. Between 1958 and 1963, for example, the share in net output of manufacturing firms with less than 200 employees declined from 20 per cent of the total to only 16 per cent, and the demise of more than half of the small companies was reported to be due to acquisition by larger rivals.[8]

During these further substantial changes in industrial concentration the major industrial groups established in the interwar period remained at the core of corporate growth. There were of course new companies attaining a leading position, such as International Computers, a firm in a rapidly expanding new industry, but the majority of large companies

[7] See Appendix 2 below. However, a more disaggregated case study approach shows that, if individual industries are considered, internal growth was an equally important source of dominance in some industries, but that this effect was balanced by internal growth of smaller firms in other industries; see, e.g., Hart, Utton and Walshe, *Mergers and Concentration in British Industry*, pp. 157–62.

[8] (Bolton) Committee of Inquiry into Small Firms, *Report* (Cmd. 4811, 1971), pp. 10–11, 59.

could trace their origins as large enterprises to the earlier period of corporate growth, and the corporations established between the wars showed a remarkable resilience. A study of the largest 100 companies of 1948, for example, found that, of the forty-eight that had 'disappeared' by 1968, nine had been nationalized, twenty-seven had been acquired by other firms and only twelve had actually regressed to a lower ranking.[9] Thus the majority of the large companies of 1948 were still among the 100 largest twenty years later, whether in their own right, or as one of the senior partners in a large amalgamated enterprise. There were, of course, significant changes in ranking amongst the top 100 and there were a number of newcomers to replace those absorbed by merger and rationalization, but the suspicion of Marshall, that 'vast joint stock companies . . . do not readily die',[10] was borne out by the continued dominance of firms like ICI, GEC, Unilever, and other large firms of long standing.

These postwar movements towards higher concentration, based to a large degree on established corporations, were in part the continuation of processes that had provided the dynamic force behind concentration increases in earlier periods. Although, in view of increasing public disquiet about monopolies, businessmen now less readily admitted monopolization or the extension of their managerial empires as a major motive of merger, these no doubt remained fundamental attractions. Again as in earlier periods of economic expansion, the rising affluence of the postwar era created larger markets both at home and abroad, and this opened up further opportunities for the division of labour and generated further economies of scale. The existence of such economies, which had previously been dismissed by sceptics, was in some degree confirmed by new studies based on engineering data. Such studies had, of course, been used by managers within large firms for a long time, but with their wider publication it could no longer reasonably be doubted that in some industries there were substantial economies of scale, which could not be achieved in the British market if there were more than a few producers.[11] In many industries in which

---

[9] K. D. George, 'The changing structure of industry in the UK', in T. M. Rybczynski (ed.), *A New Era in Competition* (Oxford, 1973).

[10] A. Marshall, *Principles of Economics*, vol. 1 (9th ed., C. W. Guillebaud, ed., 1961), p. 316.

[11] R. M. Dean and C. F. Pratten, *The Economies of Large-Scale Production in British Industry* (Cambridge, 1965). C. F. Pratten, *Economies of Scale in Manufacturing Industry* (Cambridge, 1971). A. Silberston, 'Economies of scale in theory and practice', *Economic Journal*, vol. 82, supp. (1972).

the optimal scale of production was rising, fierce competition frequently
ensued, with predictable results: the elimination of small companies,
further mergers, and the progressive concentration of output in fewer
producers. In the manufacture of television sets, for example, the number
of firms in the UK market declined from sixty in 1954 to seven by 1969.[12]
Of course, the scope for economies of scale was not confined to the
operation of large plants, and, though such economies were perhaps
the most easily measured, they may not have been the most important
ones in the minds of the managers of large companies: financial and
marketing scale economies were also considered. More generally, the
expansion of advertising (stimulated by commercial television as a new
medium) gave manufacturers an increased influence over consumer
demand, and they sometimes used this to create a more uniform consumer taste. This was sometimes complementary to the achievement
of scale-related economies on the production side. In the brewing
industry, for example, the large-scale promotion of pasteurized keg
beers (which could be transported and stored without deteriorating)
effectively widened market areas and enabled the big brewers to phase
out traditional draught beers (which needed to be brewed locally) and
to centralize production in fewer and larger breweries.[13] Scale economies in industrial research were also of increasing importance, and
Britain, alone among Western nations, devoted as high a proportion of
her resources as the United States to research and development.[14] As a
result of the increasing pressure towards larger scale enterprise, the
vogue for 'restructuring', a term now widely used to denote mergers and
the concentration of output in fewer firms, was popularized and was
strongly reminiscent of the rationalization movement of the 1920s,
both in the arguments used and in the oversimplifications to which its
less intelligent advocates succumbed.

In addition to these factors – all of them in essence a development of
tendencies inherent in earlier decades – there were also new features in
the business environment of the postwar years which accelerated the
rise in concentration. Some of the more important changes were political:
the coming to power of the Labour Party (with its more critical attitude
toward private monopolies) and the parallel change in the postwar

[12] Monopolies Commission, *Report on the Proposed Merger of Thorn Electrical
Industries Limited and Radio Rentals* (1968), p. 4.
[13] Channon, *Strategy and Structure*, p. 94. Campaign for Real Ale, *Good Beer
Guide* (Leeds, 1974), pp. 1, 96–7.
[14] M. J. Peck, 'Science and technology', in R. E. Caves (ed.), *Britain's Economic
Prospects* (1968), pp. 448–84.

Conservative Party. Already during the Second World War fears were expressed in government circles that monopolistic practices could endanger the postwar recovery, and this fear was embodied in the Labour legislation of 1948 which established the Monopolies and Restrictive Practices Commission.[15] Thereafter, stimulated by the reports of the Commission, the groundswell of opinion against restrictive business practices gathered force and a stronger deterrent, the Restrictive Practices Court, was established by the Conservative government in 1956. Whilst there were disagreements about the details, these antitrust initiatives received broad all-party support, and the Conservatives, who had once seen them as unwarrantable interferences with business autonomy, now espoused legislation, albeit somewhat uneasily, as a means of strengthening the case for private enterprise by increasing the competitive pressures within the capitalist system.[16] The Restrictive Trade Practices Act of 1956 was notable in assuming that restrictive practices were harmful unless they could be shown to be beneficial, and in the event the Court found that only 1 per cent of the agreements registered under the Act were consistent with the public interest. By 1966, 2100 registered agreements had been abandoned or terminated and there were no doubt many others which were abandoned or modified because of fears of the adverse publicity which registration might have brought.[17] Although some industries had a sufficiently oligopolistic structure for price leadership, informal agreements or secret collusion to be sufficient to maintain prices without formal restrictive agreements, the Act was successful in increasing competition in a significant number of industries.[18] At the same time, competitive forces were independently being intensified from other directions. Under the General Agreement on Tariffs and Trade (GATT), and later under the EFTA and EEC treaties, the level of tariff protection enjoyed by British industry since the 1930s was progressively reduced. With the

[15] W. H. Beveridge, *Full Employment in a Free Society* (1944), pp. 203–4. H. Dalton, *Memoirs 1931–1945* (1957), p. 447. W. T. Morgan, 'Britain's election: a debate on nationalization and cartels', *Political Science Quarterly*, vol. 61 (1946). G. C. Allen, *Monopoly and Restrictive Practices* (1968), pp. 61–4.

[16] N. Harris, *Competition and the Corporate Society: British Conservatives, the State and Industry 1945–64* (1972), pp. 221–7.

[17] Registrar of Restrictive Trading Agreements, *Report* (Cmd. 3188, 1967), p. 8. See also R. B. Stevens and B. S. Yamey, *The Restrictive Practices Court, A Study of the Judicial Process and Economic Policy* (1965).

[18] Monopolies Commission, *Parallel Pricing* (Cmd. 5330, 1973). D. Swann, D. P. O'Brien, W. P. J. Maunder and W. S. Howe, *Competition in British Industry: Restrictive Practices Legislation in Theory and Practice* (1974), pp. 144–214.

further revival of major overseas competitors, British firms no longer faced the sellers' markets of the immediately postwar years but had to meet increasing competition both at home and abroad. Competition was also intensified by a further influx of foreign (principally American) capital into the British market, which stimulated rapid growth of (often technically and managerially superior) competitors for British firms.[19] There was a widespread desire to reduce such competitive pressures within the British market, and the merger waves of these years were undoubtedly in part a response to these feelings. Such compensating mergers need not imply that there was a net decrease in competition, but certainly the initial impact of the restrictive practices legislation and tariff reductions on the level of competition was in some degree neutralized by the increased level of concentration which followed them.[20]

The intensification of competitive pressures not only produced compensating merger movements to maintain monopoly and oligopoly positions, but also, by removing some of the restraints on the natural selection of firms by competitive attrition, strengthened the motivation to agree to merger as a means of gaining access to economies of scale. In the new, relatively unprotected environment, firms which attempted to remain at an inefficiently small scale could expect to encounter greater difficulties in maintaining their market position as competition intensified. At the same time another prewar restraint on the process of adjustment to optimal scale was removed by the rise of the takeover bid. Directors of firms which had slept on their assets in the 1930s and 1940s found that in the new postwar full employment conditions, takeover bidders could acquire their shares and make substantial profits by selling off assets at their higher current prices (or sometimes by operating them more efficiently themselves). Takeover bidding became more attractive and less risky as the proportion of family dominated companies declined and as a combination of rising asset values, conservative accounting practices and dividend restraint depressed share values and improved the likely margin of profit on such a deal.[21] Government action also had its impact in this field: in

[19] J. H. Dunning, *The Role of American Investment in the British Economy* (PEP Broadsheet No. 507, 1969).

[20] Allen, *Monopoly and Restrictive Practices*, p. 98. Swann *et al.*, *Competition in British Industry*, pp. 172–8. *Economist* (11 Feb. 1961), p. 580. Cf. J. B. Heath, 'Restrictive practices and after', *Manchester School*, vol. 29 (1961), pp. 185–7. See also pp. 186–92 of this volume.

[21] See, generally: G. Bull and A. Vice, *Bid for Power* (1958); R. W. Moon, *Business Mergers and Takeover Bids* (1959); J. F. Wright, 'The capital market

particular the new accounting requirements of the 1948 Companies Act
forced companies to disclose more about their true assets and profits,
and this made it more attractive to make a bid directly to the shareholders.[22] The more aggressive takeover bidders – men like Mr Charles
Clore in the 1950s and Mr Jim Slater in the 1960s – received most
publicity, but the impact of takeover bidding extended more widely
than this. While it is true that contested bids accounted for only a
minority of actual takeovers,[23] it appears that the very threat of a bid
was in many cases now sufficient to gain the compliance of a company's
directors. As a result, the vulnerability of quoted companies to merger
was considerably increased. In the fifteen years following 1948, for
example, no less than a quarter of the companies quoted on the London
Stock Exchange were acquired by other quoted companies, and in the
ten years following 1957, when the pace of merger activity in general
also quickened, 38 per cent of quoted firms were acquired by other
quoted companies.[24] The impact of the divorce of ownership and control on creating a more fluid market in corporate control was thus
belatedly, but forcefully, established as a major pressure on the directors of industrial firms.

In the 1950s and 1960s, then, two important restraints on movements
towards higher concentration – the absence of strong competitive
forces and of takeover bids – were removed, in part as a result of
government initiatives on restrictive practices, tariffs and company
accounting. In addition to the stimulus to higher concentration caused,
paradoxically, by competition policy, governments were also more
directly active in the promotion of mergers; and, as with restrictive
practices legislation, this policy was, despite differences of approach
and emphasis, broadly bipartisan. Various factors induced governments
to abandon the *laissez-faire* approach towards manufacturing mergers
which they had adopted in the interwar years. In the first place the role of
the government as a purchaser had greatly increased as military and

and the finance of industry', in G. D. N. Worswick and P. H. Ady (eds), *The
British Economy in the Nineteen-Fifties* (Oxford, 1962), pp. 464–73.

[22] L. Hannah, 'Takeover bids in Britain before 1950', *Business History*, vol. 16
(1974), p. 75.

[23] Less than 20 per cent. See e.g.: The Panel on Takeovers and Mergers,
Report on the Year ended 31 March 1971 (1971), p. 5; H. B. Rose and G. D.
Newbould, 'The 1967 takeover boom', *Moorgate and Wall Street* (Autumn
1967), pp. 9–10.

[24] R. L. Marris, 'Incomes policy and the rate of profit in industry', *Transactions of the Manchester Statistical Society* (1964), p. 18. Department of
Trade and Industry, *Survey of Mergers 1958–68* (1970), p. 19.

welfare commitments expanded and as some major transport and energy industries were nationalized, so that an increasing number of firms depended on the government for orders and were thus obliged to respond to government pressures. In the aircraft industry, for example, where the armed forces' orders for military aircraft and the nationalized airlines' orders for civil aircraft dominated order books, the Conservative government was able in 1960 to persuade Vickers-Armstrong, English Electric, and Bristol Aeroplane to consolidate their airframe manufacturing interests into the British Aircraft Corporation.[25] This policy of intervention in private industry was considerably extended by the Labour Government of 1964–70. The earlier fears of Labour leaders that mergers and rationalization created unemployment had now given way to a feeling that larger (and, it was hoped, more productive) units would in the long run be better for employment.[26] The newly created Ministry of Technology, Department of Economic Affairs, and Industrial Reorganization Corporation (established to 'promote or assist the reorganization or development of any industry')[27] were involved in promoting larger scale units in computers, electrical engineering, motor cars, ball bearings and scientific instruments.[28] Some of the older industries also attracted the attention of government: the Shipbuilding Industry Board, for example, attempted to create viable enlarged shipbuilding groups by the merger of existing firms in private ownership.[29] More controversially, though perhaps more successfully, the British Steel Corporation in 1967 took into public ownership the fourteen largest steel makers, and planned large integrated steelworks, comparable to those appearing in Japan and the United States, which were intended eventually to replace the smaller and scattered plants which had been taken over from the private companies.[30] Although industrialists in general opposed nationalization, there were important groups of businessmen who strongly supported the government's main initiatives in promoting concentration under private ownership, and others welcomed government finance for their

---

[25] Channon, *Strategy and Structure*, pp. 110–11.
[26] It was, of course, recognized that there would be unemployment problems following mergers, but these were now felt to be problems of a transitional phase. The social costs faced by unemployed workers were also more broadly shared as a result of the redundancy payments scheme initiated by Labour.
[27] Industrial Reorganization Corporation Act (1966), S.2.1.
[28] A. Graham, 'Industrial policy', in W. Beckerman (ed.), *The Labour Government's Economic Record 1964–1970* (1972), pp. 189–91.
[29] Ibid., p. 197.
[30] Channon, *Strategy and Structure*, pp. 119–20.

ailing firms. Despite their aversion to nationalization and their abolition
of the Industrial Reorganization Corporation, the Conservative government which followed Labour in 1970 remained broadly favourable to
mergers. There were some gestures condemning Labour's support for
'lame ducks', but the policy of financial aid to industry was not abandoned.[31]

Thus government involvement in creating large companies was more
positive than that of the interwar years, when governments had been
wary of entanglements involving government finance. However, in these
merger waves, as in the past, the basic motivations in the majority of
mergers remained those inherent in modern competitive enterprise:
the classical desires to restrict competition and to achieve economies of
scale. The government's favourable attitude was generally welcomed,
and government finance was now sometimes used as a catalyst, but the
prospects of enhanced private profits remained the major driving
force. Managers were approaching the further growth of their firms
with greater confidence than in the earlier period, in part it seems because they were now less inhibited by the managerial problems which
had beset some of the earlier large companies. The progressive solution
of these problems owed much to developments in technology. The
computer and methods of operations research (both originally developed
for wartime planning) made the administration of larger scale enterprises more tractable, and the gradual build up of experience of large-
scale organization improved the capacity of large firms to undertake
further growth. Perhaps the most significant and widespread postwar
development in management was the adoption of the multidivisional
organization structure which we saw in rudimentary form in the ICI of
the 1930s. A recent study by Channon, which examined 100 large companies, found that by 1950 only 13 per cent of them had established a
multidivisional structure, whereas by 1960 the proportion had risen to
30 per cent and by 1970 to 72 per cent.[32] British subsidiaries of American companies had been among the first to adopt this structure and
British firms were often impressed by American methods and management consultants. Indeed McKinsey & Company, the consultants,
were an important force behind the introduction of the multidivisional
structure in a number of British companies.

[31] Much of the work of the Industrial Reorganization Corporation was taken
over by the Department of Trade and Industry's Industrial Development
Section. In March 1973, for example, that body offered financial aid to the
proposed BSA–Manganese Bronze merger.

[32] Channon, *Strategy and Structure*, p. 67.   [33] Ibid.

These managerial improvements not only allowed large firms to grow within their own industries but also permitted a substantial programme of diversification. In Channon's sample of large companies, for example, only 25 per cent could be considered diversified in 1950, but by 1960 the proportion had risen to 45 per cent and by 1970 to 60 per cent.[33] Much of this diversification again appears to have been the result of corporate growth by merger: between 1957 and 1968, some 39 per cent of acquisitions by quoted firms were of firms in an industry other than that of the acquirer.[34] There was also a further expansion of the overseas subsidiaries of large British manufacturing firms: whereas only 29 per cent of large firms had extensive overseas manufacturing interests in 1950, the proportion had risen to 58 per cent by 1970.[35] Yet at the same time as they were diversifying both into other industries and into foreign markets, large companies were also increasingly willing to relinquish assets under their control. Demergers – the sale of subsidiaries by companies – increased considerably: between 1969 and 1972, for example, some 24 per cent of all mergers were transfers of subsidiaries between companies (rather than acquisitions of independent companies) compared with perhaps only 2 per cent in the interwar period.[36] It appears, then, that large firms were willing, perhaps more so than their interwar forerunners, to correct 'mistakes' of diversification and to group businesses more rationally (and perhaps more monopolistically?) by reshuffling subsidiaries among themselves.[37] Thus the restructuring of industry was not, as one might have expected, inhibited by the proliferation of large diversified enterprises, for a substantial market in subsidiaries had by now been established, which was, arguably, more fluid than that under the previous régime of scattered, family ownership.

In view of the relaxation of many of the earlier managerial (and other) restraints on the growth of firms, the question naturally arises of what the future pattern of industrial concentration will be. Simple extrapolations of some of the more rapid growth rates of recent times can, of course, easily yield projections which might be considered startling. For example, one such exercise has recently suggested that three-

[34] Department of Trade and Industry, *Survey of Mergers 1958–1968*, p. 5.

[35] Channon, *Strategy and Structure*, p. 78.

[36] Recent figures from *Business, Monitor*, M7. Interwar figures calculated by the author from the merger series reported in Appendix 1 below. Some of the increase no doubt reflects the fact that in the interwar period there were in any case fewer subsidiaries and more independent companies.

[37] E. T. Penrose, *The Theory of the Growth of the Firm* (Oxford, 1959), pp. 173–82.

quarters of the private sector could be in the hands of twenty-one companies by 1985.[38] It is, however, hard to accept such extrapolations as serious predictions, though any better attempt would obviously have to be based on a fuller understanding of the parameters of concentration changes. The historical experience of merger waves provides an appropriate warning that the parameters can, in fact, change quite suddenly, and we might conclude, by analogy with the 1930s and 1940s, that a stable (or possibly even declining) level of industrial concentration would follow a rapid rise (see Chapter 9). There are of course some parallels between the present (1974) situation and the 1930s. The creation of very much larger enterprises in the 1960s may (as in the 1920s) have proceeded far enough for major gains from known economies of scale to be achieved now *within* existing enterprises rather than through further merging. This view gains some plausibility in the light of the statistics of recent changes in the sizes of plants. The smaller plants in manufacturing industry (those employing less than 200 persons) accounted for 28 per cent of manufacturing industry net output at the beginning of the upsurge of merger activity in 1958, but even after the intensive merger wave of the 1960s they still accounted for 25 per cent of net output in 1968.[39] This is reminiscent of the experience of the 1920s, where a substantial merger wave also produced very little reduction in the proportion of output produced in smaller plants, at least in the short term. If a long-term pattern similar to that experienced in the two decades that followed is repeated, then, we may expect in the coming years that the proportion of output produced in larger *plants* will increase as scale economies are embodied in new investment by the merged firms, without any increase in concentration amongst *firms* being necessary.

Yet conditions in the 1970s are significantly more conducive to further concentration than were those of the 1930s. The rise of takeover bidding, the development of multidivisional management structures, the ban on most restrictive practices, and the intensity of international competition, for example, may all now be expected to stimulate mergers, whereas in the 1930s these effects were muted. Also, as markets are now growing faster than during the depression and world war, the scope for further divisions of labour and further scale economies is also likely to continue increasing. Hence it would not be surprising if the modern rise in concentration continued indefinitely, though one might expect

[38] G. Newbould and A. Jackson, *The Receding Ideal* (Liverpool, 1972).
[39] (Bolton) Committee of Inquiry into Small Firms, *Report*, p. 59.

a more relaxed pace for a time, as the consequences of the intense merger wave of the 1960s are gradually worked out. The experience of the early 1970s provides some confirmation for this view: in 1970–3 expenditure on mergers (both in real terms and as a percentage of total investment spending) was still high, but somewhat below the average levels of the 1960s (see Appendix I).

Any future changes in the economic environment could, of course, radically alter the parameters of concentration change. Much depends, for example, on the future attitude of government to the further growth of large and powerful corporations. Already the first moves toward a hardening of government attitudes to mergers and large-scale enterprise could be seen in 1965, when Labour's Monopolies and Mergers Act strengthened the Monopolies Commission and conferred on the government powers to control mergers. By this act, the Board of Trade (later the Department of Trade and Industry) could refer to the Monopolies Commission any merger in which the assets acquired exceeded £5 million, or where the combined company would have a monopoly. This was defined, for the purpose of the act, as a situation in which one firm had a third of the market, a limit reduced to one quarter of the market by the Conservatives' Fair Trading Act of 1973. Between 1965 and 1973 the government's mergers panel considered 833 mergers which came within these definitions and referred twenty of them to the Monopolies Commission for investigation. Of these, seven were abandoned voluntarily, seven were allowed to proceed, and the six which the Commission found to be against the public interest were duly prevented.[40] This approach, a modest one compared with the more stringent US antitrust controls, was upheld by ministers of both parties who, in general, felt that the majority of mergers produced real economic benefits. They regarded mergers policy as a means of avoiding only those relatively rare cases where the benefits would not offset any social losses resulting from monopoly power.

The problems involved in this kind of judgement raised difficult and unsolved questions, but the view that ministers were right in this appraisal was not unanimously accepted and there was some pressure for more stringent controls in the American style.[41] These doubts drew

[40] J. D. Gribbin, 'The operation of the mergers panel since 1965', *Trade and Industry* (17 Jan. 1974), pp. 70–3. See also Board of Trade, *Mergers – A Guide to Board of Trade Practice* (1969).

[41] E.g. G. and P. Polanyi, 'The Fair Trading Bill and monopoly policy', *Three Banks Review* (June 1973); M. A. Crew and C. K. Rowley, 'Antitrust policy: the application of rules', *Moorgate and Wall Street* (Autumn 1971).

strength both from an ideological commitment to competitive markets and from the empirical evidence produced by research on mergers. There was, for example, some evidence that managers involved in recent mergers were concerned less with the benefits of economies of scale and more with the reduction of competitive pressure on their firms.[42] There was also increasing disillusion with the *simpliste* view, which had sometimes been accepted by politicians and businessmen, that larger scale was a prerequisite for economic efficiency: after all, the two countries with the largest plants and firms – the US and the UK – had an unimpressive growth performance compared with other countries like France, Germany and Japan.[43] This is not to say that there were no industries in Britain which had inefficiently small plants,[44] nor that the other countries could not have improved their performance even further by reaping economies of scale, but it inevitably raised doubts about the casually advanced justification for larger scale industrial units, based on unsubstantiated assertions about scale economies. The view taken by governments of evidence and arguments of this nature may, however, create a stronger political constraint on the growth of large firms in the future.

[42] Newbould, *Management and Merger Activity*, pp. 126–42.
[43] G. Bannock, *The Juggernauts* (1971). Z. A. Silberston, 'The relationship of size and efficiency', paper read to the Society of Business Economists (Cambridge, 1970). J. S. Bain, *International Differences in Industrial Structure* (New Haven, 1966).
[44] Confederation of British Industry, *Britain in Europe* (1970). G. F. Ray, 'The size of plant: a comparison', *National Institute Economic Review* (Nov. 1966), p. 63–6.

II

The upshot for welfare

Many people, who are really objecting to Capitalism as a way of life, argue
as though they were objecting to it on the ground of its inefficiency
in attaining its own objects. . . . Capitalism, wisely managed, can
probably be made more efficient for attaining economic ends
than any alternative system yet in sight, but that in itself
is in many ways extremely objectionable. Our
problem is to work out a social organization
which shall be as efficient as possible
without offending our notions of a
satisfactory way of life.

J. M. KEYNES, 'The end of laissez-faire'
(1926, as reprinted on his *Essays in
Persuasion*, 1931), pp. 320–1.

ↂ

The rise and development of the corporate economy during the present
century raises important issues both for economic analysis and for
public policy. The benefits of large corporations – and their disadvantages – have each been widely canvassed, and, though some progress in
understanding has been made as a result of this long-standing debate,
the issues raised at an early stage remain central. In the 1920s, Sir
Alfred Mond was expressing the favourable view of rapidly advancing
industrial concentration when he argued that:

> Modern mergers are not created for the purposes of creating mono-
> polies or for inflating prices. They are created for the purpose of
> realizing the best economic results which both capital and labour
> will share to the best advantage. They enable varieties of industries
> to form an insurance against fluctuations of markets and prices in
> individual products. . . . Amalgamations mean progress, economy,
> strength, prosperity.[1]

Nonetheless, his case did not lack contemporary censure, and critics
with equally simplistic recipes for economic welfare soon replied. The
classic liberal economists' condemnation of monopoly power remained
the basis of Francis Hirst's unfavourable judgement:

[1] Sir Alfred Mond, *Industry and Politics* (1927), p. 217.

From the standpoint of general utility there is all the difference
between the promoter of a real new enterprise and the promoter
of a combination or amalgamation. The former is calculated to
increase wealth; the latter is rather likely to diminish it. The former
is good for employment, the latter is likely to reduce it. The former
increases the good things of the world and multiplies the conveni-
ences of life. The latter aims at restricting them and so increasing
their cost. One is addition, the other subtraction. One enlarges the
world's resources and enriches the consumer by giving him something new; the other exploits him by establishing a monopoly and
so forcing him to pay higher prices or to pay the old prices for
inferior articles.[2]

The dilemma posed by the contrasting views of Mond and Hirst has
persisted. Indeed, with the further advance of the corporate economy
the dilemma has been posed with increasing urgency, and provokes
often heated debate among economists and policy makers on the nature
and advantages of large corporations.

The problem raised is a fundamental one. It arises because an increase in concentration, whether it results from mergers or from the
internal growth of large corporations, is (*pace* Hirst) neither entirely
'addition' nor unrelieved 'subtraction'; rather, it contains something
of both. The corporate economy has on the one hand created new
possibilities for promoting economic growth, yet it has also established
an industrial structure with a potential for severely restricting the level
of economic welfare and the equitable distribution of income. This
coexistence of costs and benefits (which lay at the root of the disagree-
ment between Hirst and Mond) poses no less difficult a dilemma for
modern governments. As we have seen, the Labour government of
1964–70 with one voice charged the Monopolies and Mergers Commission with controlling concentrated industries, yet with another encouraged industrial concentration through the Industrial Reorganization Corporation. Its opponents were quick to point to a contradiction
between these policies, but their criticism appears to bypass the
essence of the dilemma which welfare judgements in this area must
confront. In point of fact, if it were possible to separate the beneficial
from the baneful effects it would be perfectly sensible to promote the
one whilst curtailing the other. (This is not to say that the policies
in fact pursued by the Monopolies Commission and the Industrial

[2] F. W. Hirst, *The Stock Exchange* (1932), p. 222.

Reorganization Corporation were consistent; the point is simply that they were not *necessarily* inconsistent.)

In practice it is usually difficult for a government agency to assess fully and quantitatively either the benefits or the costs of any monopolistic situation currently existing. *A fortiori*, for the historical case it will not be possible for us to assess the entire impact of the rise of the corporate economy in terms of net benefits. This would require a detailed knowledge of the cost and demand conditions faced by entrepreneurs over the whole period, and such information is simply not available. Nor is it possible to argue that, since 'the extent to which combinations make for efficiency is a matter on which industrialists are the first experts',[3] we can take the continuing expansion of large companies as *prima facie* evidence that the benefits exceeded the costs in the long run. This view rests on assumptions which, upon further examination, prove untenable. First, a company enlarged by merger may survive because its size and power confer a purely private benefit which corresponds to a social loss: this would arise, for example, in mergers undertaken to achieve monopoly powers. The survival and extension of the corporate economy may, then, in itself merely indicate imperfections of competition rather than the supposed benefits of large corporations. A second assumption on which the technique of inferring welfare gains from the surviving structure of industry rests is that, at the moment of observation, the economy is in a position which reflects the optimal organization of production. In fact, however, what we observe may simply be one point in a process of adjustment. This process may involve lags between initial managerial errors and their eventual correction by compensating structural change. Although, as we have seen, mergers were *ex ante* motivated by a number of potential benefits, *ex post* these benefits may have been realized only imperfectly if at all. Indeed, perhaps the most consistent conclusion of research on mergers over the last fifty years is the finding that a large number of them fall short of the hopes and intentions of their promoters. We have already noted the failure of the multi-firm mergers of the turn of the century and the belated managerial response as the diseconomies of rapid growth were perceived. Yet when the trend to higher levels of industrial concentration was resumed in the 1920s, errors and innefficiencies of a new kind appeared. The financial results of these mergers were frequently below those promised by promoters and there were spectacular failures of firms promoted both by unscrupulous financiers

[3] D. H. MacGregor, in R. Liefmann, *Cartels, Concerns and Trusts* (1932), p. xi.

and by honest (but misguided) industrialists who had fallen prey to the jargon of 'rationalization' without having the managerial skills to implement its substance.[4]

Another common problem in this early stage of corporate growth arose from sour personal relations between the directors of acquired companies who remained part of the management team. The directors of formerly independent firms (especially owner-managers of family firms) often found it difficult to accept that loss of personal power and independence of action which was usually required if economies of scale and coordination were to be released through a merger. Yet, despite their difficulties, such mergers often had an advantage which was denied to some partners in the later waves of takeovers. In many recent bid situations the deal has often been hurriedly and forcibly consummated: there has thus been little rational and informed discussion of the industrial logic of the decision by the boards of the merging companies and only a minimal opportunity to plan its implementation. The results can be seen in the serious difficulties of post-merger integration experienced in recent merger waves, and the modern statistical evidence on the economic performance of merging companies inspires no greater confidence than the historical record. Singh, for example, in a recent study, reported that almost two-thirds of the companies involved in takeovers experienced a decline in profitability in the year of merger and in the following year.[5] Of course it can be argued that takeover bids may have compensating advantages, as a scourge to managements using company assets inefficiently, but it is hard to resist the view that the price paid for this stimulus is high.

The persistence of this pattern of shortcomings (which can also be

---

[4] *Economist* (26 July 1930), p. 182. However, the suggestion of F. R. Jervis (*The Economics of Mergers* (1971), p. 51) that the majority of interwar mergers were unsuccessful is exaggerated, and his examples of capital reductions by merging companies could be paralleled in these years of depression by examples of non-merging companies forced into similar capital reductions, suggesting that some of his 'failures' should be seen as responding rationally to the general depression of these years rather than as failures in the true sense.

[5] A. Singh, *Takeovers: Their Relevance for the Stock Market and the Theory of the Firm* (Cambridge, 1971), ch. 7. The decline in profitability is measured relative to the profitability of non-merging firms in the same industry. See also M. A. Utton, 'On measuring the effects of industrial mergers', *Scottish Journal of Political Economy*, vol. 21 (1974). To the extent that an effective merger involves much new investment, a decline in short-run profitability is desirable, but case histories suggest that organizational difficulties have also contributed to the decline.

observed in American mergers over a long period)[6] suggests that
managers consistently make errors in assessing the profitability of
mergers. This is plausible in view of recent findings that firms proposing
mergers do not, in general, do so on the basis of a full and informed
assessment of the costs and potential savings which might be available.[7]
City opinion also has been increasingly critical of undiscriminating
merger proposals from corporate managements.[8] So pervasive is the
evidence of the failure of merging companies' managements to live
up to their own predictions of success that some general explanation
of the phenomenon seems to be required. It seems plausible that, in
general, managers have a prejudice in favour of direct, 'rational' and
'scientific' management in their own firms as against the more imper-
sonal and subjectively more uncertain operations of the market. Hence
they will welcome the replacement of existing relationships with other
firms through a market, as a means of coordinating their interrelated
economic activities, by the organization of these activities within the
boundaries of one firm. We have seen that this kind of managerial bias
towards management rather than market played its part in the ideologies of 'rationalization' in the 1920s and of 'restructuring' in the
1960s. We need speculate no further, but there remains a strong *prima
facie* case for believing that such a *déformation professionelle* exists and
that it has often clouded judgement and induced managers to believe
ex ante that they could more effectively manage an enlarged enterprise
than *ex post* appears to have been the case. The social costs of their ill-
advised decisions may have been substantial.

To point to the short-term difficulties of mergers is not, however,
to suggest that in the long run the benefits of increasing concentration
have not been realized. Indeed, the historical evidence is equally strong
that economies of scale have been achieved over a wide range of British
industries and that, as many of these economies are internal to the firm,
they could only have been realized through the expansion of large undertakings. The long-run tendency for the average size of both plants and

---

[6] A. S. Dewing, 'A statistical test of the success of consolidations', *Quarterly
Journal of Economics*, vol. 36 (1921). S. Livermore, 'The success of industrial
mergers', *Quarterly Journal of Economics*, vol. 50 (1935). S. R. Reid, *Mergers,
Managers and the Economy* (New York, 1968), pp. 28, 44–6, 63–4, 91–5.
M. Gort and T. F. Hogarty, 'New evidence on mergers', *Journal of Law and
Economics*, vol. 13 (1970).

[7] Monopolies and Mergers Commission, *Guest Keen & Nettlefolds Ltd and
Birfield Ltd. A Report on the Merger* (1967), p. 30. G. D. Newbould, *Management and Merger Activity* (Liverpool, 1970), pp. 113–16.

[8] The *Guardian* (11 Feb. 1974), p. 20.

firms to rise is suggestive of the widespread existence of increasing
returns to scale or at least of constant returns; and the standardization
and mechanization of manufacturing production, notably through the
growth of assembly line production in large factory complexes, has undoubtedly been an important source of improving productivity in
manufacturing industry. Recent research into engineering production
functions confirms that in many industries the economies of large-scale
production are so great that if there were more than a few firms in the
market in these industries it would be impossible to reap all these
benefits of scale. Further, there is evidence that, at least before the
merger wave of the 1960s, firms in some industries were still not large
enough to reap all the advantages of scale and that, if managerial constraints had permitted, there could have been a social benefit to even
higher levels of concentration.[9]

The benefits of large scale are often considered exclusively in terms
of the growth of plant size, this being an area where they have been
much in evidence. This emphasis may, however, be misleading, for the
rate of increase in the size of plants has been somewhat slower than the
rate of increase in the size of firms. Prais has suggested, for example,
that whereas the share of the top 100 *plants* in manufacturing net output remained constant at around 11 per cent between 1935 and 1963,
the share of the top 100 *firms* rose rapidly – from 24 per cent to as much
as 38 per cent over the same period.[10] This suggests that the advantages
of large-scale enterprise, if they exist, do not lie exclusively in their
ability to operate plants of larger scale. The wider benefits which are
potentially available from within firms (rather than simply within
plants) are implicit in the analysis of the causes of increasing concentration of earlier chapters. One important case is the relative efficiency
of the capital market and of large firms in allocating investment resources. If, as we have argued, some large firms in the interwar years
were able to assess future returns on investment projects more accurately
than the institutions of the stock market, because of their greater
knowledge of production technology and marketing, then the channel-
ling of investment funds through them will have had favourable effects
on the quality of investment decisions. More recently, however, it
might be argued that the ill-informed and discontinuous disciplines
and often downright fraudulent promotions of the interwar stock
market have been ameliorated. With the development of institutional

[9] R. Pryke, *Public Enterprise in Practice* (1971), pp. 350–2. See also p. 167
above. [10] S. J. Prais, unpublished paper.

investment, more stringent statutory accounting requirements and the takeover bid, the stock exchange is perhaps better able both to assess new investment projects and to maintain an efficiency check on past ones. That the stock market has indeed overcome some of its former handicaps is suggested by the finding of recent research that equity-financed earnings of quoted companies in general are, in fact, marginally more profitable than earnings on projects financed by quoted firms themselves through reinvested profits.[11] This is consistent with the view that the stock market now imposes a useful discipline on firms by requiring them to achieve a reasonable rate of return. It does not, however, necessarily imply that large firms are becoming inherently more inefficient, since the equity-financed investments are themselves increasingly channelled through large firms. These firms do, then, appear to retain important advantages as allocators of funds and as analysts of capital projects, as in the past.

A further important advantage of large companies (and one which has been increasing over time) is the promotion of invention and innovation. The heavy commitments of research and development undertaken in modern science-based industries such as chemicals and electronics could only have been shouldered by large firms, and these firms have also proved adept at importing new technology from abroad and at developing and marketing the inventions of others. A régime of pure competition between small firms is unlikely to produce the right conditions and incentives for successful invention and innovation. By contrast, large firms with substantial market shares not only have extensive resources to devote to research but are also better able to internalize the externalities in producing knowledge, by extending its use to their existing products and markets. This is not to say that the individual inventor and the innovating small company are extinct – indeed the evidence is that in many industries they are still very much alive[12] – but rather that in much of science-based industry the substantial financial commitments, the risks and uncertainties of the research process, and the range of technical and marketing expertise necessary for successful innovation give the large-scale organization a head start over its competitors. The flow of new products to the market, from ICI's invention of polythene in 1933 to more recent innovations

[11] G. Whittington, 'The profitability of retained earnings', *Review of Economics and Statistics*, vol. 65 (1972).
[12] J. Jewkes, D. Sawers and R. Stillerman, *The Sources of Invention* (1958), pp. 91–126.

such as carbon fibres, could hardly have been as rapid (and in some cases may not have occurred at all) without the intermediation of large research organizations of the kind we have described.[13]

If they had been able to review, retrospectively, the advantages of large scale in this way, many of those involved in creating the modern corporate economy would, if the expressed hopes of Sir Alfred Mond are a guide, have had little hesitation in seeing the improved growth performance of the British economy in the interwar years, and again in the postwar period,[14] as an important consequence of the changes in the size and nature of firms. These changes were, as we have seen, conspicuous by their absence before the First World War but particularly marked during later decades when the rate of economic growth also accelerated. A favourable view of the corporate economy is therefore compatible with the evidence of improvement in Britain's growth performance over time. The productivity improvements and investment decisions lying behind this growth are, of course, the result of a variety of factors, in addition to the managerial and organizational changes which we have been considering, but most studies of British growth agree that economies of scale, technical innovation and organizational efficiency have, historically, made important contributions to economic growth. There are conceptual difficulties and serious problems of measurement which inhibit any convincing quantitative assessment of the overall importance of these factors relative to other sources of economic growth (for example, improvements in education and in government macroeconomic policy), but that the benefits were real and substantial is hardly open to doubt.[15] However, against these benefits must be juxtaposed the social cost of the monopolies which appear to be an integral part of the modern corporate economy: the

[13] This is not to say that the continued expansion of large firms can be justified on scale grounds. Many companies are now sufficiently large to gain access to scale economies in R and D without further expansion, as the Monopolies Commissions recognized in its *Report on the Proposed Merger Between Boots Pure Drug Co. Ltd and Glaxo Ltd* (1971). See also F. M. Scherer, *Industrial Market Structure and Economic Performance* (Chicago, 1970), pp. 352–62.

[14] R. C. O. Matthews, 'Some aspects of post-war growth in the British economy in relation to historical experience', *Transactions of the Manchester Statistical Society* (1964).

[15] For an attempt to measure the relative importance of the various sources of growth in the UK, see E. F. Denison, *Why Growth Rates Differ* (Washington, 1967), pp. 314–15. Denison suggests that Britain enjoyed fewer of the benefits which might be linked with the corporate economy than other European countries. Whether that is due to the particular nature of British corporations or to some general failing in British society is a moot point.

large firms whose rise we have described are not only absolutely large but large relative to individual markets, which they are thus able to dominate.[16] Whether the efficiency gains of increasing concentration are passed on to the consumer, and whether industry will continue to be stimulated to operate efficiently, will crucially depend on the behaviour of firms which is induced by this new, and possibly less competitive structure. We must therefore examine more closely the impact of the growth of corporations on the relative domains of monopoly and competition in the economy.

The widely held view that monopoly power has been steadily increasing (and that at present levels it imposes a serious burden of inefficiency on the community) rests, in part, on the now familiar evidence of the substantial and continuing rise in industrial concentration over the economy as a whole and in individual industries. This, it is inferred, has inflated prices and profits for firms in monopolized or oligopolized industries: a not altogether implausible inference, since it is widely recognized and admitted by business groups themselves that obtaining monopoly power has always been an important aim of mergers and the expansion of large firms.[17] To identify high levels of concentration with monopoly may, however, be misleading. There is, of course, *some* relationship between the exercise of monopoly power and the level of concentration; where there are few firms or where one large firm dominates the others, output is more likely to be restricted and prices will be higher than in an industry in which there are many small firms each with a tiny share of the market. In practice, however, there will be many other variables affecting the degree of competition in an industry, of which the threat of entry by new firms, competition from imports, the speed of technical change, the level of advertising and the extent of spare capacity are perhaps the more measurable (though there are others, less tractable to quantitative analysis but perhaps none the less real, such as the pure animal spirits of the competitive instinct and the much-vaunted commitment to competition as a business ideal). Since many of these variables have clearly been changing over time, it would be improper to conclude without taking explicit account of them that the well-documented increases in concentration have caused a net increase in monopolistic behaviour.

16 M. A. Utton, 'Aggregate versus market concentration', *Economic Journal*, vol. 84 (1974), pp. 150–5.
17 British Electrical and Allied Manufacturers' Association, *The Electrical Industry in Great Britain* (1929), p. 194. Newbould, *Management and Merger Activity*, p. 139.

In some respects the monopolistic tendencies suggested by increasing concentration have actually been reinforced by complementary forces. Modern corporations, for example, have at their disposal sophisticated techniques of market research and an accumulation of experience of a variety of pricing policies, and are thus better equipped than their forerunners for maximizing monopoly revenues through price discrimination. Also monopolists and oligopolists have, over time, extended their knowledge of the entry-preventing price – that is, the price which will enable them to maximize their profit without giving new competitors sufficient incentive to enter their market. Other techniques of raising barriers to entry have also been developed: fighting companies to annihilate local competition; the filing of restrictive patents or vexatious law suits; schemes for exclusive collective rebates; these have all been successfully aimed at stifling competition. The reality of dangers of this kind, inherent in highly concentrated industries, is attested by the many postwar reports of the Monopolies Commission. This unique source of information on the improving techniques of monopolization leads one to suspect that only the unimaginative businessman would today reject a merger offer on the ground (advanced in the 1920s) that 'a new business will spring up for every one that is merged'.[18] It is true that some of the more predatory techniques have been effectively controlled by the action of the Restrictive Practices Court and of the government's monitoring system, but many remain, and there are alternative routes to strengthening market power that have so far escaped government intervention. Product differentiation through branding and advertising, for example, has reinforced oligopolies. At the beginning of this century, the advertising industry was of negligible size, but by the 1960s advertising expenditure had grown to almost 3 per cent of consumer expenditure. Similar developments have occurred in branding: before the First World War some 95 per cent of dry grocery goods were taken from bulk and broken into small lots by the retailer, whilst at present almost all are pre-packaged and branded. The high degree of consumer loyalty which these products boast often reflects contrasting product 'image' rather than real differences in quality. Despite the social waste involved in much branding and advertising of this kind, government action has so far been con-

[18] Anon., 'Why we *did not* join the amalgamation', *Business* (Jan. 1928), p. 15. On restrictive practices see, generally, Monopolies and Restrictive Practices Commission, *Collective Discrimination. A Report on Exclusive Dealing, Collective Boycotts, Aggregated Rebates and other Discriminatory Trade Practices* (1955).

fined to reducing advertising expenditure by detergent manufacturers, a limited policy but one which has successfully provided a competitive stimulus by inducing new entry from firms previously unable to enter because of heavy advertising barriers. Elsewhere, however, companies remain free to maintain entry barriers through advertising.[19]

However, before concluding that competition has been gradually expiring as high concentration, entry barriers and product differentiation have been established over a wide area of the market, we should recall that over the same period the market has itself changed, with effects running counter to these forces. What the twentieth century has seen (but what national concentration figures fail to reveal) is a transformation of many local oligopolies into regional, national or international oligopolies. Competition may not, then, have been reduced if the average size of markets has been increasing over time. Given the trend reduction in transport costs (partly a result of road–rail competition and, more recently, of the introduction of containers) it does appear that effective market areas both at home and abroad have indeed been greatly expanded. Moreover, with urbanization and the spread of universal education and a national press and television, demand in these enlarged markets has become increasingly standardized. Thus while many of the consumer goods industries in the late nineteenth century served mainly local markets with local tastes, the more concentrated industries of today (for example, the producers of domestic electrical goods) serve a mainly national and standardized market. In these circumstances it is, of course, possible that a modern firm, even though it has a larger share of the national market in a given product, will face more competition than its predecessors in the last century who enjoyed protected local markets. Moreover, many firms compete in an international market, not only by exporting a high proportion of their output, but also in the sense that a list of their competitors in the home market would properly include overseas producers and importers with large market shares. At the turn of the century, under a régime of free trade, the British economy was perhaps more open to *potential* foreign competition than it is today, but, as other countries have rapidly industrialized, the number of *actual* competitors has multiplied. Moreover, international agreements, such as GATT

19 *Economist* (22 Oct. 1938), pp. 156–7. N. Kaldor and R. Silverman, *A Statistical Analysis of Advertising Expenditure and the Revenue of the Press* (Cambridge, 1948). Evely and Little, *Concentration in British Industry*, p. 140. P. Doyle, 'Economic aspects of advertising: a survey', *Economic Journal*, vol. 78 (1968).

and the EFTA and EEC free trade areas, have more recently greatly reduced the level of tariff protection for British industry. It would, then, be rash to proclaim the rapid increase in concentration in the 1960s as conclusive evidence of increasing monopolization in the home market without taking explicit account of the sevenfold increase in imports of finished manufactures over the same period. Of course the size of market can also decline and when it does so it may work in concert with the forces of concentration to increase monopoly power. The abandonment by Britain of free trade in 1932, for example, reinforced the impact of the merger waves of the 1920s in reducing competition, and it was not until the 1950s that the trend to higher tariff barriers was reversed. British withdrawal from the Common Market, if not accompanied by tariff reductions elsewhere, would also have comparable effects in reducing market size and hence of inhibiting competition.

The size of the market is determined not only by the geographical spread of the sales area but also, importantly, by the extent of competition between products or processes which are close substitutes. Here again there is reason to believe that the effective market for the average product has been growing, in a manner unreflected in the national concentration data,[20] as new industries and a diversification of demand have together increased inter-product and inter-industry competition. Synthetic substitutes for natural materials have been introduced and new products increasingly compete with old: margarine with butter, rayon and nylon with cotton and wool, synthetic chemicals with natural ones, radio transmission with telecommunications cables, reinforced concrete and plastics with structural steel and wood. Large companies have at various times tried to bring their inter-product competition under control, but with little success, except where complete dominance went to one side, as with the victory of electric light bulbs over gas mantles. Even in industries where product innovations have been less in evidence, the standardization of existing products – a tendency which was particularly strong in the engineering and electrical fields where there were significant technical advantages to standardization – has increased the substitutability of the products of firms which were once protected from competition by a degree of product differentiation. On the demand side also, the elasticity of substitution between average

[20] Even where concentration data for individual industries (rather than the economy as a whole) are available, the classification is often based on the technical characteristics of the production process rather than on the cross-elasticities of demand between products which would be conceptually more appropriate.

products has probably increased in the course of the present century as rising income levels have confronted consumers with a wider spectrum of potential purchases. A television set, a washing machine and a holiday abroad are perhaps more real alternatives than were foodstuffs, clothing and housing at the turn of the century, when they took up a much larger part of the typical family budget than is now the case. From many angles, then, there appears to be good reason for believing that the effective market size which a typical producer faces has been greatly expanded in the course of the twentieth century.

In view of the many and varied determinants of monopoly power, its precise measurement is elusive, but, even if we could show that the factors making for monopoly overrode countervailing trends such as increases in market size, and that perfect competition was replaced by monopoly in many individual markets, the effects on economic welfare of these changes could not be determined unambiguously. It is a well-known theoretical proposition that if there is monopoly *anywhere* in the economy (as there surely has been at all times) an increase in monopoly elsewhere will not necessarily decrease welfare and may, indeed, increase it.[21] A familiar case in point has occurred in the distribution sector, as retail chains and department store groups have greatly expanded their share of consumer sales at the expense of the small shopkeeper. Because of their highly concentrated market shares, these large groups have been able to use their bargaining power with producers and this 'countervailing oligopsony' has to some extent neutralized the monopoly powers of manufacturing firms. It is, of course, possible that both parties in a situation of this kind will perceive that it is in their interests to collude and to raise prices to the consumer,[22] but where there is substantial competition in one sector – as there still is between retailing groups in large towns – their countervailing power has been a significant restraint on corporations.

The increase in the level of concentration which we have observed in manufacturing industry was in part a response to precisely these factors which were increasing competitive pressures on firms. To picture merger waves as motivated by a desire to suppress a long-

---

[21] R. G. Lipsey and K. Lancaster, 'The general theory of second best', *Review of Economic Studies*, vol. 24 (1956–7). I. M. D. Little, *A Critique of Welfare Economics* (2nd ed., 1957), p. 163.

[22] Large-scale buyers frequently seem more interested in preventing oligopolistic competitors from receiving more favourable prices than with securing an absolutely lower price; see, e.g., W. J. Reader, *Imperial Chemical Industries: A History*, vol. 1 (1970), pp. 372–3.

established régime of perfect competition would, then, be misleading: it would be rather more accurate to picture them as attempts to counteract increases in competitiveness which were occurring for independent reasons. The competitive threat of countervailing oligopsony from large retail chains, for example, has been moderated for some manufacturers through the acquisition of retail outlets: film makers, for example, control the majority of cinemas, and brewers the majority of public houses. Another threat arises where economies of scale have led in the short run to overproduction and unremunerative prices, and again firms have been anxious to neutralize the resulting competitive pressures. Within individual firms, knowledge of the existence of scale economies created concern about the financial losses which would ensue in conditions of over capacity, and hence induced agreements with competitors or a full merger of interests to ensure that prices covered costs. Also, widening markets, new entry and legislation against agreements between firms have all added to these pressures, so it becomes clear that the substantial merger movements of the 1920s and 1960s had to neutralize important long-run developments of competitive forces before they could have any net positive effect in increasing monopoly power.

Under closer examination, then, the initially plausible inference that the substantial measured increase in industrial concentration in the British economy over the twentieth century indicates a substantial increase in monopoly power appears less clear cut. Had other things remained equal, we would perhaps be justified in treating higher concentration levels as synonymous with enhanced monopoly power, but other things were evidently very far from equal. If we are to be able to make a valid statement about changes in the degree of competition and monopoly, then, we will need to devise an alternative approach to the problem. One option would be to attempt to weigh the various factors which we have isolated as important determinants of competitiveness and of monopolistic behaviour. Indeed the reader may perhaps have already formed his own opinion of the relative weight to attach to each of the considerations which we have adduced to modify the initial inference from concentration data alone. If so, he will be able to reach a conclusion on the direction of change, but it must be recognized that reasonable disagreements on the weighting of individual factors would cast this conclusion in doubt.[23] An alternative, and conceptually

[23] For an extreme view that competition has substantially increased, see Industrial Policy Group. *The Growth of Competition* (1970). The data on

neater, strategy would be to measure changes in monopoly power directly, for example by measuring the divergence of prices and marginal costs for a sample of products over time. However, the absence of relevant data rules this out as a realistic option.[24] We are, then, left with no clear knowledge of the direction or rate of change in the degree of monopoly in Britain during the present century.

Those who find the restrained agnosticism of this conclusion unpalatable may take comfort from the reflection that it need not rule out judgements on whether individual industries are likely to suffer from monopolistic distortions and inefficiency. Indeed recent research suggests that areas ripe for government investigation on grounds of monopoly can usefully be pinpointed by means of appropriately qualified concentration data.[25] What our agnostic conclusion does suggest, however, is that we should approach sceptically any overblown claims about the supposed euthanasia of competitive capitalism and its replacement by monopoly capitalism. Curiously, this view has appeal both as a diagnosis of the ills of the modern corporate economy by its critics[26] and as a vehicle for the hope of its friends that their moral discomfiture at the harsher aspects of competition can be purged as competitiveness is progressively replaced by a spirit of cooperation and service.[27] The protagonists in this debate would do well to remem-

---

industrial concentration which the authors adduce to support their conclusion are wrong, but the major difficulty in accepting their case is the absence of a firm basis for their implicit weighting of the various determinants of competition.

[24] The rate of profit in industry might be suggested as a proxy measure of monopoly power, its long-run decline being taken as an indication of increasing competition. However, there are problems both of conception and measurement in this conclusion: profits are the result of other forces besides monopoly power, and the accounting data on which the decline in the rate of profit has been measured use definitions of profit which are unsatisfactory in this context. Using a parallel approach, W. E. G. Salter in his *Productivity and Technical Change* (Cambridge, 1966, pp. 158–60) concludes that most increases in productivity between 1924 and 1950 were not captured in profits but were passed on to the consumer in the form of lower prices.

[25] P. E. Hart, M. A. Utton and G. Walshe, *Mergers and Concentration in British Industry* (Cambridge, 1973). G. Walshe, *Recent Trends in Monopoly in Great Britain* (Cambridge, 1974).

[26] Notably by Marxist critics of capitalism, but also – strange bedfellows – by believers in the market system who hope for a regeneration of capitalism through the promotion of competition. The latter view is aggressively stated in the various publications of the Institute of Economic Affairs.

[27] E.g. *The Times* in a leader headlined 'The silent social revolution' (25 Mar. 1935) opined that 'Competition has disappeared over a large portion of the industrial field and the motive of profit, the mainspring of the capitalist

ber that the evidence for their underlying assumption of the euthanasia of competition is far from clear cut.

If, as appears from this examination of the determinants of competition and its welfare implications, the considerations are so diverse and depend on the particular conditions faced in each industry at different points in time, then it follows that a government policy aiming at maximizing welfare should be *ad hoc* rather than general. There have been advocates of a more thoroughgoing policy, but prescriptions such as the breaking up of monopolies or a ban on all concentration-increasing mergers ignore the difficulties raised by the absence of any clear presumption that the benefits of such a policy would exceed the costs. Even if the benefits of a more stringent policy could be shown to be positive (and on this there is clearly room for scepticism), they would in any case probably be small compared with the more substantial gains available from policies designed directly to increase technical efficiency and promote economic growth.[28] This has led some economists to postulate a relationship not only between competition and allocative efficiency but between competition and other organizational and dynamic aspects of overall efficiency. It has been suggested, for example, that organizational slack, the choice by managers of a quiet life, and a disinclination to innovate, will all be more common in an economy which lacks the strong performance incentives created by the insistent pressures of competition.[29] On this view the stimulus given to efficiency by promoting a greater degree of competition in the economy will exceed the traditionally recognized, and probably small, gains derived merely from reducing the allocative distortions of monopoly. (Parenthetically, it can be noted that this view of the benefits of promoting competition accords more closely to popularly accepted notions: the idea of the competitive 'spur' to greater efforts and more efficient management is a familiar one.)

---

system, has a decreasing importance. . . . Service rather than profit will undoubtedly be the keynote of the age into which we are passing.' Similarly inflated suggestions have been made at regular intervals since.

[28] A. C. Harberger, 'Monopoly and resource allocation', *American Economic Review Papers and Proceedings*, vol. 44 (1954). But cf.: G. Stigler, 'The statistics of monopoly and merger', *Journal of Political Economy*, vol. 64 (1956), pp. 33–5; H. Leibenstein, 'Allocative efficiency *vs* "X-efficiency" ', *American Economic Review*, vol. 56 (1966).

[29] W. S. Comanor and H. Leibenstein, 'Allocative efficiency, X-efficiency, and the measurement of welfare losses', *Economica*, vol. 36 (1969). R. M. Cyert and K. D. George, 'Competition, growth and efficiency', *Economic Journal*, vol. 79 (1969).

However, while this supplementary notion of the potential benefits of competition (and the corresponding dangers of monopoly) is the more popular, it is both less well founded in its logical structure and less well substantiated empirically than the more orthodox propositions of analytical welfare economics about allocative efficiency. Economics lacks a proven theory of the relationship between industrial structure and competitive behaviour on the one hand and the dynamic phenomena of improving efficiency and overall growth on the other.[30] The problem arises in part because 'competitive' has a meaning separate from and independent of that which it takes as the antonym of monopoly; it can be a description of a state of rivalry and striving rather than of a market structure.[31] The implications for economic efficiency of this second meaning have been worked out only informally and unsystematically. 'Competitiveness' in this sense may be the most important variable in improving overall efficiency, but this is not proven, and, even if it were proven, 'competitiveness' could be achieved without the large population of firms which is normally considered to be a precondition of competition in its first sense. Indeed an industry with a small number of large firms may exhibit more rivalry than a highly disaggregated industry, because rivalry is partly a function of the awareness by participants of the activities and achievements of the rivals. Paradoxically, this awareness (which is vital to constructive rivalry) is more likely to flourish in an oligopoly than in a perfectly competitive world of price takers, a paradox which is reflected in the seemingly perverse business usage of phrases such as 'healthy competition' in praise of an industry practising output and price controls.[32]

A further problem in interpreting the impact of competition on efficiency is that rivalry, being a subtle mixture of cooperation with competition, can be organized within the firm as well as stimulated by the market. Thus an increase in concentration may actually increase the spirit of rivalry among managers. The rationalization movement of the 1920s, for example, emphasized emulation of the excellent and insisted on the need to develop greater comparability of results internally as a means of measuring and achieving that excellence. Large firms reacted to this by developing new post-merger management structures which replaced coordination through market competition

[30] R. L. Marris, 'Why economics needs a theory of the firm', *Economic Journal*, vol. 82 (1972).

[31] P. J. McNulty, 'Economic theory and the meaning of competition', *Quarterly Journal of Economics*, vol. 82 (1968).

[32] E.g. A. W. Rather, *Is Britain Decadent?* (1931), p. 173.

by coordination within the firm by its management. Moreover in this process they also attempted to recreate under ideal conditions some of the clear-cut decision rules which the competitive system of resource allocation supposedly gave, with the result that it became possible to turn the initial argument for competition on its head so that it became an argument for consolidation:

> It is questionable [one enthusiast for rationalization wrote] whether the competitive system provides scope for such direct and fruitful competition as this pitting of experts one against the other in their own field. Costs and processes are kept top secret, and no direct comparison is possible between the costs and efficiency of rival firms. . . . Organized competition [i.e. within the firm] pits like with like and measures their comparative efficiency with precision; the free play of the competitive system [i.e. the market] confers its rewards and punishments indiscriminately. Organized competition and the encouragement of initiative and enterprise are essential to the success of large-scale organization.[33]

To the extent that the large multidivisional firms whose rise we have described were successful in developing this kind of constructive rivalry, the social benefits of 'competitiveness' will have been replicated internally to the firm without an absolute need for a competitive market stimulus.

Even when the more nebulous benefits of the competitive stimulus to efficiency are taken into account, then, there is still no general case for believing that the industrial developments which we have described have been fundamentally harmful. Nevertheless while a competitive régime of production and exchange cannot be shown to have exclusive net advantages over a more concentrated one, there is evidence that the elimination of the competitive spur can lead to the hardening of the economic arteries, if compensating controls internal to the firm (or external ones like the threat of takeover) are not developed. Before the restrictive trade practices legislation of the 1950s, there were numerous examples in large manufacturing organizations of inefficiency bolstered by competitive restrictions in their markets.[34] More recently, Turner & Newall has been pinpointed by the Monopolies Commission as a company whose growth initially produced benefits from economies of scale and improved coordination but which in the long run lost its

[33] E. M. H. Lloyd, *Experiments in State Control* (Oxford, 1924), p. 361.
[34] G. C. Allen, 'An aspect of industrial reorganization', *Economic Journal*, vol. 55 (1945).

impetus to improved performance and by the 1960s was simply stag-
nating.[35] It is not implausible that the efficiency of other large firms
does suffer from the absence of a competitive stimulus: there is evidence,
for example, that large multidivisional firms do not use checks on
internal efficiency, such as interdivisional market pricing, which would
enable them to monitor their own performance more effectively.[36]
However, neither the government nor the market economy itself is
entirely bereft of means to counteract these dangers. The threat of a
takeover bid or nationalization, and direct pressure on the firm from
institutional investors (or from the Department of Industry acting
under powers derived from the monopolies legislation), can all, in some
degree, make up for the long-term disadvantages which society might
otherwise suffer from entrenched monopolies. Clearly the further the
corporate economy advances the greater will be the corresponding
dangers and the stronger the case for reinforcing and extending these
powers.

In recent years governments have, of course, acted on these principles and strengthened their monopoly policy, but they have also
developed their relationships with large firms in a more positive sense,
as a means of achieving wider policy aims. Given the importance of
large firms in the economy – we have seen that the largest 100 of them
together account for almost one half of manufacturing output – it is
hardly surprising that they have become an important adjunct of
government. Indeed the convenience of using them as vehicles for
government economic policies no doubt in good measure explains why
British governments have been reluctant to adopt a more thoroughgoing 'antitrust' policy. The indicative planning initiatives of the 1960s,
for example, depended crucially on corporations for providing information and responding to targets. The need to exempt the Economic
Development Committees from restrictive practices legislation in
order to encourage the informed planning of investment projects by
large firms was also recognized in that period.[37] Again, in the 1970s,
attempts to control inflation relied heavily on the cooperation of the

[35] Monopolies Commission, *Report on the Asbestos and Asbestos Cement Industries* (1972).
[36] D. F. Channon, *The Strategy and Structure of British Enterprise* (1973), pp.
196–217. There is no very clear evidence on whether large firms in general
are more, or less, profitable than smaller ones. For a discussion of recent data,
see: J. M. Samuels and D. T. Smyth, 'Profits, variability of profits and firm
size', *Economica*, vol. 35 (1968); T. E. Chester, 'Large organizations – their
role in the UK economy', *National Westminster Bank Review* (Aug. 1971,)
pp. 32–4.   [37] Restrictive Practices Act (1968).

Confederation of British Industry and on the monitoring of the prices of the largest 200 companies. Although the results were not universally applauded, it is perhaps significant that under the counterinflation controls of 1973, price increases by smaller and medium-sized companies (which were not under close surveillance) were over twice as high as those by the large companies (which were).[38] The opening up of these wider policy options has, then, entered – and perhaps rightly entered – into the balance of economic costs and benefits which governments have seen in the corporate economy.

An assessment of the impact of the corporate economy which focuses solely on the economic costs and benefits is, however, inevitably incomplete, for corporations exercise a pervasive influence not only on economic but on political and social life also.[39] Wider considerations of political and social welfare have always been important in public assessment of modern industrial tendencies. The Labour Party, for example, has been willing to concede much of the economic case for concentration: 'In terms of efficiency these vast centralized concerns are often, but by no means always, justified. . . .' But they have seen serious problems of political control: 'The greatest single problem of modern democracy is how to ensure that the handful of men who control these great concentrations of power can be responsive and responsible to the nation.'[40] The policy of nationalization (which had its origin in a desire to reduce inequalities of wealth by taking direct control of industrial assets) has thus become adapted, in an age when ownership of assets is increasingly divorced from managerial control, to the new problems of the corporate economy. In this sense more recent programmes of nationalization can be seen both as a response to and a possible further development of the corporate economy. On the one hand nationalization is greatly facilitated when control of industry is already highly centralized. The 'nationalization' of many hundreds of firms into one organization has in the past created problems similar to those encountered by earlier multi-firm mergers in the private sector.[41] More recent

---

[38] The *Guardian* (5 Oct. 1973), p. 1. The inference that there were substantial gains in restraining inflation from the ability to monitor large companies seems plausible, though some of the discrepancy may be due to contrasts in the cost conditions faced by small and large firms.

[39] J. K. Galbraith, *The New Industrial State* (1967). J. A. Schumpeter, *Capitalism, Socialism and Democracy* (1943).

[40] Labour Party, *Signposts for the Sixties* (1961), p. 10.

[41] Most of the nationalized firms of the 1940s took ten years to achieve real gains in managerial efficiency; see Pryke, *Public Enterprise in Practice*, pp. 19–35.

nationalizations, by contrast, have proceeded more smoothly, perhaps in part because it is managerially less difficult to amalgamate fewer companies. This was the case, for example, in the more concentrated steel industry, nationalized in 1967. At the same time, fears of monopolization or of financial weakness in the private sector of industry have led to greater willingness of governments, whatever their political hue,[42] to investigate nationalization as a means of achieving the benefits of large scale without its disadvantages. Public ownership may, then, in the longer term, have a substantial and increasing part to play in the corporate economy.

This is not, however, the place to predict the future, nor, indeed, to prescribe for it. Prophets of Western capitalism have in the past foreseen far-reaching economic, social and political changes as the inevitable result of corporate growth, but their predictions have been characterized more by inflated claims than by precision. Like the historians of the corporate economy (and no doubt for similar reasons) its prophets have been divided between optimists and pessimists and each has appealed to a wide range of considerations to justify their stance. Individual liberty in a pluralist society has been ascribed to the dispersion of control of that society's assets: hence rule by the few has been forecast as the likely result of a more unequal distribution of managerial power. The progressive integration of big business into the social and political élite has raised fears of a vulgar plutocracy of 'bastard capitalism';[43] and there are fears also that advertising by monopolistic corporations will continue to degrade language and logic as it has in the past. More generally, there is evidence that, throughout the social scale, alienation is more acute in a society based upon large-scale enterprise, and a wide range of opinion now views the working conditions and working relationships necessary for the achievement of scale economies as fundamentally dehumanizing.[44] Critics of these developments offer the alternative of small-scale or cooperative enterprise as a means of building on altruism and the cooperative spirit to achieve collective goals. On the other hand, there are those who believe that these goals

[42] It was a Conservative government that nationalized the aero engine division of the Rolls Royce Company. Initially this was because of financial failure, but increasingly state control of the company has been justified on the grounds that only the state can shoulder the risks of research and development in this industry.

[43] R. Samuel, 'Bastard capitalism', in E. P. Thompson (ed.), Out of Apathy (1960), pp. 19–55.

[44] For an excellent review of the sociological literature on these subjects, see J. Child, The Business Enterprise in Modern Industrial Society (1969).

can be more efficiently and satisfyingly achieved within the corporate framework: the values of professionalism, cooperation, altruism, deliberation and excellence can, it is argued, be pursued within small *working* groups, even in an economy in which control of *assets* is concentrated. It is natural that in assessing the effects of the profound modern changes in the structure of industry and in anticipating future trends, businessmen have stressed the benefits rather than the costs, and critics have focused attention on the losses rather than the gains. The historian can record their views, though he can hardly produce a judgement more definitive than theirs, given the complexity of the corporate economy and its diverse and intricate social and political implications.[45]

Nevertheless, if precise quantification of both the economic and social welfare implications of the corporate economy is elusive, the central feature of the growth of the large-scale corporation as a significant institution of the modern economy is unmistakable. With the concentration of capital, and the organization of economic life within large firms, the economy moved away from individualist market mechanisms and moved closer towards that centralized economic control which Keynes and the liberals had advocated as a means of preserving the capitalist system and socialists had seen as a stage in its overthrow. In place of the disaggregated market economy of the nineteenth century came the mixed economy of large corporations and an unequivocal movement towards greater coordination by non-market forces within a bureaucratic structure of decision making. As the corporate economy advanced, men espoused its cause, partly on ideal grounds and partly on grounds of efficiency, yet they have often seen it fall short of their expectations. Critics and prophets can clarify the mistakes of the past and the future choices available, but even in the clearer perspectives of the postwar years the value of our inherited economic structure has been a matter of fundamental political debate, and it will no doubt remain so. If we have succeeded in explaining how and why the transformation which gave rise to this debate took place, we will have achieved our major purpose, but the wider questions raised by this transformation lie beyond the scope of this work and in the realm of the choice of an alternative future not by past generations but by the present one.

[45] G. Hannah and Kay, *Concentration in Modern Industry.*

Statistical appendices

ॐ

The reason of the thing is not to be
enquired after, till you are sure the thing
itself be so. We commonly are at *what's
the reason of it?* before we are sure
of the thing.

JOHN SELDEN, *Table Talk* (1689), p. cxxi.

ॐ

Appendix 1

Merger waves in United Kingdom manufacturing industry 1880-1973

ಣಲ

Ideally merger statistics are compiled by a government agency with the aid of statutory merger disclosure requirements and an efficient financial press. Unfortunately for the historian of past merger movements, the late development of antitrust in this country leaves him with no reliable official statistics of acquisitions until the period from 1954 onwards. Moreover, in contrast to the position in the United States, where private enterprise was eager to fill this gap, British economists appear to have had only sporadic interest in mergers before they received the incentive of government policy criticism, which has since turned merger analysis into a flourishing academic industry. Between the excellent private enterprise study of the late Victorian merger waves by the Fabian civil servant Henry Macrosty,[1] and the assumption by the state of responsibility for the compilation of merger statistics, there is no very reliable tabulation of merger activity. J. M. Rees transcribes (sometimes inaccurately) the limited number of reports of the committees on trusts which were published between 1919 and 1921;[2] while P. Fitzgerald and A. F. Lucas limit their coverage to a few of the major mergers of the interwar period and to price associations.[3] It was therefore necessary to construct an entirely new series for merger activity in Britain between 1880 and 1954. This task is described in detail elsewhere,[4] and readers are referred to these works for an account of the sources, assembly and classification of merger data for the years

[1] H. W. Macrosty, *The Trust Movement in British Industry* (1907).
[2] J. M. Rees, *Trusts in British Industry 1914–1921* (1921).
[3] P. Fitzgerald, *Industrial Combination in England* (1927). A. F. Lucas, *Industrial Reconstruction and the Control of Competition* (1937).
[4] L. Hannah, 'Mergers in British manufacturing industry, 1880–1918', *Oxford Economic Papers*, vol. 26 (1974). L. Hannah, *The Political Economy of Mergers in Manufacturing Industry in Britain between the Wars* (unpublished DPhil thesis, Oxford, 1972).

prior to 1940 and for fuller descriptions of the merger waves of these years, disaggregated by industry, by type, and by size categories. The purposes of this appendix are: first, to summarize the main outlines of the methods of compilation used; and secondly, to link the series, in so far as that is practicable, to the government statistics which begin in 1954 and which are more fully described in various government publications.[5]

Except where otherwise indicated, the term merger is taken to include both consolidations (in which a new holding company is formed to acquire the constituent companies) and acquisitions of one company by another. The treatment of both kinds of merger is standardized: if company A acquires company B, one 'firm disappearance' is recorded: if A and B merge to form a new company, C, one firm also disappears, the value of the smaller of A and B being taken in this case as the value of the 'disappearing' firm. Sales of subsidiaries between independent companies are also counted as 'firm disappearances' by merger. In general, only firms with significant assets in manufacturing industry and operating principally in the United Kingdom are included. A merger is defined as the acquisition by one company of more than 50 per cent of the voting power of another, and thus conforms to the criterion of control adopted in studies of concentration based on the census of production. Prior to 1954 the date of merger is taken to be the date of the decision (or of the announcement of intention) to merge, but in the official statistics it may be either the accounting year in which the merger took effect or (from 1969) the date of the announcement of the consummation of the merger.

The statistical series of merger activity presented in Table A.1 below is derived variously from business and industrial histories, year books, company accounts and reports in the financial press, and it would be rash to suppose that such heterogeneous sources had produced a series which was comparable in coverage at all points in time. An asterisk has been placed against points in the table where the major breaks in series occur and the reader is referred to the notes on subperiods below for an indication of the varying definitions and of the difficulty of

[5] 'Acquisitions and amalgamations of quoted companies 1954–1961', *Economic Trends* (Apr. 1963). 'Acquisitions and mergers of companies', in *Trade and Industry* (published quarterly). *Business Monitor*, Miscellaneous Series, M7 (published quarterly). Department of Trade and Industry, *A Survey of Mergers 1958–1968* (1970). I am indebted to Mr J. L. Walker of the Department of Trade and Industry (Economics and Statistics Division) for supplementary advice on the use of the government statistics.

effecting linkages. Broadly speaking, the modern data are probably the fuller and more reliable since they have been compiled in an environment created by the more stringent disclosure requirements of the postwar years and by greatly improved financial reporting services. The historical series, by contrast, relies variously for its raw data on mention in a less efficient financial press, on the commissioning of business histories by particular firms, on the interest of scholars studying particular industries in chronicling mergers, and on the historical investigations of the Monopolies Commission. It would be surprising if the mists of history had not obscured some past merger activity which under modern conditions would have been recorded. Even the current series, compiled in more advantageous contemporary circumstances, is by no means complete.[6] The merger activity shown in our table for any date is, then, only the tip of an iceberg, though, unlike the case of the iceberg, we cannot be sure of a constant relationship between the visible and submerged portions. A further problem of comparability arises where it has not been possible to maintain constant definitions of the population to which the statistics relate. Between 1954 and 1968, for example, the population includes acquisitions of foreign companies by UK quoted companies, but excludes some domestic acquisitions by non-quoted companies, whereas for 1880-1953 and 1969-73 this position is reversed. It follows from this, and other changes in definitions noted below, that comparisons of the level of merger activity in widely separated years will be subject to potentially large margins of error. The series has been used in Chapters 7, 9 and 10 to establish a general presumption that the merger waves of 1919-30 were somewhat larger than those of the 1930s and 1940s (where the data would, if anything, probably be biased against this conclusion). It has also been taken to show that the merger-intensive 1920s were in some respects comparable to the 1950s and 1960s (where other evidence also suggests that the comparison is not wholly fanciful). Any attempt to use this imperfectly linked series for more precise analysis or for broader historical comparisons would, however, be hazardous, if it did not take account of the difficulties involved.

In the historical data, and to a lesser extent in recent data, the number series (column 1 in Table A.1) is likely to be more accurate than the value series (column 2), which rests on a mix of prices paid, market values and nominal values, and in the early years, on some arbitrary

[6] (Bolton) Committee of Inquiry into Small Firms, *Report* (Cmd. 4811, 1971), p. 10.

assumptions about unknown values. It is not immediately obvious whether numbers of firms disappearing by merger or their values (gross expenditure on acquisitions) are the more appropriate measure of the volume of merger activity. The preference for a value index to be seen in much of the recent merger literature derives from a feeling that the index should reflect a higher figure if two companies with a market valuation of millions of pounds merge than if the merging companies are valued at only a few thousand pounds. However, while this point has some force in studies of the effects of mergers on concentration or on the growth of large firms, where the sizes of the partners are important, it is not unambiguous even for that purpose. An example will serve to show this ambiguity. If there are three companies, X, Y and Z, with capitalizations respectively of £4 million, £2 million and £1 million, the value series will show discrepant results for very similar economic occurrences. For example, if X acquires first Y and then Z, £3 million will be shown as disappearing; while if Z first acquires Y and then X acquires $(Y + Z)$, £5 million in all will be shown as disappearing.

By contrast, the inadequacies of a number series can easily be overstated. If ten firms each worth £100,000 are acquired it is arguable that this should be reflected by a higher index number than the acquisition of one firm worth £1 million, for it does involve a correspondingly greater reduction in the number of independent decision making units in the economy and a correspondingly greater number of decisions to merge. The number index also has the advantage of being in constant terms at different points in time, though the index of merger values at constant share prices (column 3 in Table A.1) compensates for this weakness in the value series.[7] (Share prices are a more appropriate deflator than retail prices since they reflect, in addition to the general change in the price level, the changing national stock of assets of which merging assets form a part. In comparing the 1920s with the 1960s, for example, we wish to take account of the fact that in the 1920s the capital stock in manufacturing industry was perhaps only two-fifths and retail prices only one third of their level in the 1960s: deflation by a share price index provides a rough approximation to this desired dual correction.) One advantage of using the value series is that, unlike the number series, it is not greatly altered by the addition of large numbers of small firms, and thus comparisons between widely separated years will be

[7] Moodies' share price index was used to correct the values for 1919–73 to 1961 share prices. For the period 1880–1918, the London and Cambridge Economic Service share price index, spliced to Moodies at 1919, was used.

less hazardous for the value series since most large mergers (but a changing proportion of the smaller mergers) are included at all dates.

A further variant of the value series (again, however, subject to the same weaknesses as the raw value data) is shown in column 4 of the table. This indicates the proportional contribution of external growth (i.e. growth by merger) to the total growth, both internal and external, of manufacturing firms in each year. Using gross domestic fixed capital formation as an indicator of internal growth,[8] merger activity can be expressed as a proportion of 'total investment expenditure' (i.e. internal and external growth combined). The economic meaning which can be attached to this proportional measure is, of course, limited since 'total investment expenditure' is an artefact of the financial data, not a fixed fund of real resources. While individual firms may have a choice between spending their surplus cash flow on the acquisition of existing firms and spending it on investment in new plant and buildings, the physical constraints on new investment in the economy as a whole differ from the constraints on the overall level of merger activity. For some purposes, however, the proportional contribution of merger activity to the growth of firms may be the more relevant measure.[9]

The following notes indicate the major characteristics of the various series in the six major subperiods between 1880 and 1973.

1880–1918

The only series in existence for this period, based on Macrosty, was found to be inadequate, and a new series (tripling Macrosty's coverage of mergers and extending the series from 1907 to 1918) was constructed, using business histories, industrial studies and contemporary reference works. A check in the investors' press suggested that the use of this as a further source would have improved coverage only slightly, at least until the later years of the period. The value series for this period is an extremely tenuous estimate: approximate figures for the price paid were available for only 44 per cent of the firm disappearances by merger, and the values of the remaining firm disappearances were arbitrarily

---

[8] Gross Domestic Fixed Capital Formation in UK Manufacturing and Building at current prices for 1920–38 and 1949–65, from C. H. Feinstein, *National Income Expenditure and Output of the United Kingdom 1855–1965* (Cambridge, 1972), p. T93; and, for more recent years, from Central Statistical Office, *National Income and Expenditure* (1972), p. 66, and *Economic Trends*, No. 246 (Apr. 1974).

[9] E. T. Penrose, *The Theory of the Growth of the Firm* (Oxford, 1959), pp. 241–2.

assumed to average one quarter of the average known values of firm disappearances in the same decade.

1919–38

Two major series were constructed, one based on business and industrial histories, and the other on mergers reported in the *Investors Chronicle*. This method of compilation enabled corrections for omissions to be made, for although the probabilities of a merger occurring in each series were not known, the total finite population of mergers (of which the two series were independently drawn samples) could be estimated. The figures thus derived (some 41 per cent higher on average than the raw series of directly observed mergers) are only lower bound estimates of the total numbers of mergers.[10] The values in column 2 for these years were grossed up from known values. The values of 63 per cent of the firm disappearances in the raw series were known, and the range of values presented in column 2 was estimated on the assumption that the unknown values were between one tenth and one quarter of the average known values of the same year.[11] The constant price value series in column 3 is also presented as a range. However, the proportion of total investment expenditure accounted for by mergers (column 4) appears as a single value, the middle of the estimated range of values being chosen for these purposes – implying, that is, an assumption that unknown values averaged one seventh of known values. In one year, 1926, the value series is, perhaps disproportionately, affected by the largest merger of the interwar years, the formation of Imperial Chemical Industries, in which three firms valued at £36 million (about one half of the total values estimated for 1926) disappeared.

1939–53

A series for these years has not previously been available, but it was possible to construct one by using a method akin to that used in the earlier series. In this case, however, statistics of mergers reported in the financial press have not been collected, and the method of correction for omissions using probability theory was thus not available. Instead the raw series derived from business histories and similar sources was grossed up by a multiple based on its relationship to the final estimate of merger activity for the interwar period. The series for this period is

---

[10] Hannah, *Political Economy of Mergers*, p. 133.
[11] For a justification, see ibid., pp. 136–7.

therefore somewhat less reliable than in the preceding or succeeding periods, since the year-to-year fluctuations are likely to be exaggerated by the narrower sample of mergers from which the figures are estimated. However, there is no reason to believe that the level of merger activity indicated for the period as a whole is misleading. A comparison both with the government statistics for 1954 onwards and with mergers reported in the financial press confirmed that the link at 1953–4 can be made without serious distortion of the level. The numbers of firm disappearances calculated for 1954, 1955 and 1956 by the method used for 1939–53 were 208, 190 and 198 respectively. The corresponding numbers reported in the official series were 197, 196 and 181. A supplementary check of merger activity reported in the financial press confirmed that, as this comparison indicates, the government figures for these years were slightly on the low side.

Because of the paucity of information on values in this period, total values were not estimated for 1940–8. The values for 1939 are calculated on the same basis as those for 1919–38. The values for 1949–53 are from an independent series based on company accounts[12] and, like the values for the later 1950s, they relate to a variety of accounting years and to acquisitions by quoted companies. Since these values are derived from a different source, we cannot be certain that they correspond directly to the numbers of firm disappearances shown in parallel in column 1.

1954–9

This series is from the government statistics produced by the Board of Trade in its analysis of company accounts. For a variety of reasons the figures for this period diverge most from the idealized definitions laid down in our preamble. The series is based on the accounts of companies whose financial years end within the twelve months to the 5 April following the year indicated. It thus relates to accounting, not calendar, years. It includes only acquisitions of independent companies by companies in the sample and transfers of subsidiaries between companies in the sample. Consolidations of independent companies (i.e. mergers in which a new holding company acquires two or more existing firms) are excluded. Two large consolidations, had they been included, would have increased the value index considerably: in the Yorkshire Imperial Metals merger of 1958 the smaller company (Yorkshire Copper Works) had net assets disappearing of £8 million, and in the Unigate merger of

[12] National Institute of Economic and Social Research, *Company Income and Finance 1949–53* (1956), p. 23.

1959 the smaller company (Cow & Gate) had net assets disappearing of £12·9 million. The series has a contrary tendency to overestimate the level of domestic merger activity by including acquisitions by UK quoted companies of foreign companies, which have been excluded in the years prior to 1949.

1960–8

The series is similar to that used for 1954–9 in that it is derived from company accounts, but there were two important changes in the procedure. First, from 1961 there was a net reduction of some 400 small companies (with assets of under £0·5 million or with income of £50,000 or below) in the Board of Trade's population. This probably reduced measured merger activity by 6·6 per cent relative to the earlier years, though the coverage was again widened somewhat in 1964 and 1968. Secondly, the improved treatment of consolidations, introduced in 1971 by the Department of Trade and Industry, has been applied retrospectively to these statistics: consolidations are therefore included, the large firm being considered the acquirer and the smaller the acquired, thus standardizing their treatment with that of acquisitions. Transfers of subsidiaries between companies, and acquisitions and consolidations of independent companies (including acquisitions by UK companies of foreign companies), are included.

The values recorded in 1968 were at a historically unprecedented level. Two large mergers in that year, British Leyland Motors and GEC – English Electric, accounted for £732 million, or 44 per cent of the merger values recorded.

1969–73

The change in the basis of the official statistics in 1969 from company accounts to reports in the financial press enabled a number of improvements to be made. In particular, it was possible to date mergers by the calendar year in which the transaction was finalized rather than by the accounting year, to exclude acquisitions of foreign companies by UK quoted companies, and to widen coverage to include more non-quoted companies. Some quoted UK companies operating mainly overseas were also added to the population so that their UK acquisitions could now be included. Unfortunately, however, this change actually reduced coverage of manufacturing mergers considered as a whole, and this

weakens the link between 1968 and 1969. We have figures for only one
later year calculated on the same basis as the period 1960-8: in 1969,
621 firms valued at £724 million disappeared in mergers according to the
earlier method, compared with the 481 firms valued at £722 million
shown on the revised basis in our table. It might reasonably be inferred
that a large number of small acquisitions, mentioned in company
accounts but not in the financial press, were eliminated by the revision of
method, and that the significance of the break in the value series is
less than that in the number series.

TABLE A.1 *Merger activity in UK manufacturing industry, 1880–1973*

	Number of firm disappearances by merger	Values (at current prices) of firm disappearances (£ million)	Values (at 1961 share prices) of firm disappearances (£ million)	Merger values as a proportion of total investment expenditure (%)
1880	4	0·1	1	–
1	1	0	0	–
2	6	1·7	21	–
3	6	0·2	3	–
4	1	0	0	–
5	8	0·3	4	–
6	11	0·3	4	–
7	21	0·5	7	–
8	101	5·3	75	–
9	48	1·7	21	–
1890	92	8·9	111	–
1	35	0·8	10	–
2	24	0·7	9	–
3	11	0·2	3	–
4	17	0·4	5	–
5	32	0·9	10	–
6	69	5·8	53	–
7	83	4·3	36	–
8	151	8·3	70	–
9	255	11·5	94	–
1900	244	21·9	181	–
1	49	7·0	61	–
2	76	9·6	85	–

TABLE A.I (continued)

Number of firm disappearances by merger	Values (at current prices) of firm disappearances (£ million)	Values (at 1961 share prices) of firm disappearances (£ million)	Merger values as a proportion of total investment expenditure (%)
3	53	4·2	38	–
4	32	1·5	15	–
5	39	2·5	23	–
6	34	2·1	19	–
7	42	1·9	17	–
8	18	1·9	18	–
9	72	2·7	26	–
1910	38	9·9	89	–
1	63	8·1	69	–
2	58	5·5	47	–
3	31	3·2	28	–
4	32	2·9	26	–
5	44	4·8	45	–
6	43	4·0	33	–
7	41	7·8	57	–
8	112*	26·2*	158*	–
9	288*	89–101	446–508*	–
1920	336	59–67	317–359	29
1	78	14–16	110–125	11
2	67	11–14	73–93	16
3	124	22–28	121–149	30
4	129	12–15	62–77	18
5	116	45–49	202–220	36
6	153	70–77	301–332	51
7	180	38–44	154–175	37
8	270	44–51	155–180	39
9	431	45–50	159–176	38
1930	158	28–31	120–136	30
1	101	16–19	88–105	25
2	86	9–10	49–57	16
3	92	14–17	67–82	23
4	121	16–18	66–75	18
5	187	18–20	67–75	20
6	274	30–37	98–124	26

TABLE A.I (*continued*)

	Number of firm disappearances by merger	Values (at current prices) of firm disappearances (£ million)	Values (at 1961 share prices) of firm disappearances (£ million)	Merger values as a proportion of total investment expenditure (%)
7	174	19–23	63–75	14
8	127★	21–26	84–104	18
9	94★	13–17	57–74	–
1940	36	–	–	–
1	44	–	–	–
2	23	–	–	–
3	65	–	–	–
4	85	–	–	–
5	88	–	–	–
6	99	–	–	–
7	125	–	–	–
8	109	–	–	–
9	104	15	48	4
1950	49	10	31	2
1	78	4	10	1
2	185	9	28	1
3	122★	42★	125★	7★
4	197★	91★	203★	13★
5	196	67	127	8
6	181	120	247	12
7	223	109	210	10
8	251	100	206	9
9	385★	245★	320★	21★
1960	513★	313★	324★	22★
1	486	479	479	26
2	479	302	322	20
3	583	290	271	20
4	700	432	385	24
5	668	440	421	22
6	572	443	426	21
7	525	756	677	32
8	631★	1666★	1112★	49★
9	478★	716★	488★	26★

H

TABLE A.I (*continued*)

	Number of firm disappearances by merger	Values (at current prices) of firm disappearances (£ million)	Values (at 1961 share prices) of firm disappearances (£ million)	Merger values as a proportion of total investment expenditure (%)
1970	426	668	545	23
1	424	372	265	14
2	576	1292	672	37
3	610	458	273	15

Summary of merger activity by decade[13]

1880–9	207	10	136	–
1890–9	769	42	401	–
1900–9	659	55	483	–
1910–9	750	161–173	998–1060	–
1920–9	1884	360–411	1654–1886	32
1930–9	1414	184–218	759–907	21†
1940–9	778	–	–	
1950–9	1867	797	1507	10
1960–9	5635	5837	4906	28
1970–3	2036	2790	1755	23

Notes:
– Data not available
*Major break in series
† 1930–8 only

[13] In comparing merger activity in different decades regard should be paid to the strong *caveats* in the introduction to this appendix.

Appendix 2

Industrial concentration in
the United Kingdom
1907-70

ಬಬ

The measurement of industrial concentration poses many perplexing problems of both a theoretical and an empirical nature. The theoretical issues have been discussed in some detail elsewhere.[1] Hence the aim of this appendix is limited to placing some of the major empirical contributions in the existing literature on industrial concentration into perspective, by linking previous research and data collected in the course of the present study into a description of trends in concentration in the United Kingdom over the twentieth century as a whole. It is intended to amplify the statistical analysis of industrial concentration in Chapters 7 and 10, and also introduces examples of indices of concentration which are more appealing technically than those employed there.

The simplest and most readily understood measure of concentration, and one we have used extensively in earlier chapters, is the concentration ratio: the share of the largest n firms in the total sales or output of an industry. Conventionally, when measuring concentration in manufacturing industry as a whole, n is taken as the top fifty or 100 firms. Since data on the sales or output of individual firms are not available for most years, such varied measures of size as profits, assets, market valuation of capital, and employment have also been used. Unfortunately, this diversity of approach vitiates direct comparisons between many studies of concentration, but it is nonetheless possible, by various methods of interpolation, using a range of assumptions, to construct a reasonably full statistical series of the share of the largest 100 firms in manufacturing net output over the period between 1909 and 1970.[2] The

---

[1] E.g. L. Hannah and J. A. Kay, *Concentration in Modern Industry: Theory, Measurement and the UK Experience* (forthcoming 1976).

[2] I am indebted to Dr S. J. Prais for helpful discussion of the problems of interpolation involved. See also his 'A new look at the growth of industrial concentration', *Oxford Economic Papers*, vol. 26 (1974).

full series is shown in Table A.2 below, and has also been plotted in graph form in Figure 7.1 (see p. 105).

TABLE A.2 *The share of the largest 100 firms in manufacturing net output, 1909–70*

1909	15%	1948	21%
1919	17%	1953	26%
1924	21%	1958	33%
1930	26%	1963	38%
1935	23%	1968	42%
1939	23%	1970	45%

Source: See text.

In calculating the shares shown in Table A.2 four different methods were used, depending on the best information available for the various dates.

Method 1 (1909, 1924)

P. E. Hart's estimates of the shares of the largest fifty firms in total profits in manufacturing in 1908–10, 1924 and 1938[3] were corrected, using more accurate data on manufacturing profits than were available at the time of his original study.[4] By assuming that the change in the share of the largest 100 firms in net output was proportional to the change in the share of the largest fifty in profits, it was possible to derive estimates of the share of the largest 100 firms in net output in 1909 and 1924, using a method suggested by S. J. Prais for linking the data with their known share in 1935.[5]

Method 2 (1919, 1930, 1939)

Data on the share of the largest 100 firms in the market valuations of the capital of a range of firms in manufacturing industry were readily available for 1919 and 1930.[6] On the assumption that the increase in con-

[3] P. E. Hart, 'Business concentration in the United Kingdom', *Journal of the Royal Statistical Society*, Series A, vol. 123 (1960), p. 52.
[4] G. D. N. Worswick and D. G. Tipping, *Profits in the British Economy 1909–1938* (Oxford, 1967), p. 44. P. E. Hart, *Studies in Profit, Business Saving and Investment in the United Kingdom 1920–1962*, vol. 1 (1965), p. 21.
[5] S. J. Prais, 'Notes on the measurement of the share of the largest 100 companies' (Appendix A of forthcoming study).
[6] Hannah and Kay, *Concentration in Modern Industry*.

centration between 1919 and 1930 was linear,[7] their share in 1924 can also be estimated. Further, by reconstructing Hart and Prais's data for 1939 and excluding non-manufacturing enterprises,[8] an estimate of the share of the largest 100 firms in market values on an approximately equivalent basis could also be derived for that year. Allowing for changes in the coverage of the populations at the various dates, and on the further assumption that changes in market valuation were proportional to changes in net output, estimates of the share of the largest 100 firms in net output for 1919, 1930 and 1939 can be derived by linking this market value series, at 1924, to the estimate of the share of the largest 100 firms in net output derived by method 1.

Method 3 (1935, 1958, 1963, 1968)

The estimates derived by this method are the most recent and the firmest. They rely directly on information in the *Census of Production* on the net output of large enterprises, in the case of 1958, 1963 and 1968.[9] For 1935 the census tabulation by Leak and Maizels was adjusted by S. J. Prais to take account of non-manufacturing enterprises.

Method 4 (1948, 1953, 1970)

From the census series derived by method 3, the estimates for these years were made by S. J. Prais by interpolation on the basis of financial accounts.[10]

In each of the four basic methods there are undoubtedly errors in estimation. The *Census of Production* data on net output are subject to margins of error, and these are magnified by the assumptions necessary in the interpolations for years in which *Census of Production* data could not be used, whether because no census was taken in that year or because

[7] This assumption is supported by the observed incidence of merger activity in 1919–24 relative to 1924–30.

[8] P. E. Hart and S. J. Prais, 'The analysis of business concentration: statistical approach', *Journal of the Royal Statistical Society*, series A, vol. 119 (1956), p. 154. The sizes of the larger firms were calculated directly, but for the smaller firms the middle points of the size categories published by Hart and Prais were used. It was not possible to add a range of unquoted firms to the 1939 population of quoted firms in order to make it directly comparable with the 1930 data.

[9] The census results for 1968 were provisional at the time of writing.

[10] S. J. Prais, 'Notes on the measurement of the share of the largest 100 companies'. The data for these years are provisional at the time of writing, and, in particular, 1948 may be revised upwards.

the returns have not yet been analysed in a manner which facilitates this kind of calculation. It is unlikely either that the share of the top fifty firms moved in the same way as that of the top 100, or that changes in the market values or profits were proportional to changes in net output. The possibility of errors arising from such assumptions renders comparisons between widely separated years particularly hazardous since the effect of the errors may be cumulative. Even in comparisons between adjacent years small movements in concentration may be accounted for entirely by errors in the data or by faulty assumptions in the interpolations. Until the earlier *Census of Production* returns have been more carefully analysed, however, it is unlikely to be possible to produce a more acceptable description of changes in manufacturing industry as a whole.

A further weakness of the approach exemplified in Table A.2 is that the choice of the share of the top 100 is a matter of computational convenience rather than of economic significance. It is possible that changes in concentration, as alternatively measured by, say, the share of the top ten or the top 1000 firms, would be different, and, if this were so, it would be a matter of some interest. Ideally, then, what is required is a measure of the concentration ratio for all values of n, rather than for one of them: a measure of the whole distribution of firm sizes. Unfortunately it is difficult, especially in historical periods, to estimate the sizes of every firm in the economy, as such a measure would ideally require. Hence studies which have attempted a more generalized measure of concentration have usually been limited to quoted companies (and thus to the upper ranges of the size distribution of firms) for which estimates of size based on stock market valuations of capital can be made. P. E. Hart and S. J. Prais, for example, have estimated the sizes of quoted companies in manufacturing and distribution for the years 1907, 1924, 1939, 1950 and 1955.[11] However, there are difficulties in interpreting their conclusions on concentration since they used a measure of concentration – the variance of the logarithms of firms' sizes – which is inappropriate and potentially misleading.[12] None the less it proved possible to reconstruct their population of firms by calculating the sizes

11 Hart and Prais, 'The analysis of business concentration', p. 154. P. E. Hart, 'Concentration in the United Kingdom', in H. Arndt (ed.), *Die Konzentration in der Wirtschaft* (Berlin, 1971), appendix. They also estimated the sizes of firms in 1885 and 1896, but since the quoted sector was of only minor significance in that period, the results are of little interest for the measurement of overall concentration.
12 Hannah and Kay, *Concentration in Modern Industry*.

of the top dozen or so firms in each year from data published in the *Stock Exchange Daily Official List* and by using the midpoints of the size classes published by Hart and Prais for the smaller firms (in which deviations from the midpoints of size classes would not significantly affect the result). The results are shown in Table A.3. This uses a comprehensive numbers-equivalent measure of concentration which avoids the pitfalls of the variance of logarithms. This measure, K, which was originally suggested by J. A. Kay, is given by the expression

$$K = \left(\sum_i s_i^{\alpha} \right)^{\frac{1}{1-\alpha}}$$

where s_i is the market share of the ith firm and α is the elasticity of market power with respect to firm size.[13] The first column of Table A.3 indicates the range of values of α used. When α takes a low value (in practice $\alpha = 0.6$ is the lowest value used) changes among smaller firms in the lower part of the distribution will have a stronger impact on the index than when α takes a higher value. A high value of α (in practice $\alpha = 2.5$ is the highest used), on the other hand, will give great weight to the largest firms in the distribution.[14] The table thus supplies a more comprehensive measure than either a single concentration ratio or a single elasticity parameter.

The population of quoted firms (on which Table A.3 is based) accounts for a portion of manufacturing industry which rises over time, as more companies gain quotations. Thus it is important to compare constant samples of companies rather than the whole distributions for each year.[15] The appropriate pairwise comparisons of constant samples can be readily distinguished in the table since they lie between the thicker vertical rules. From these comparisons it appears that, for all the relevant range of elasticities ($\alpha = 0.6$ to $\alpha = 2.5$), concentration within this population rose in 1907–24 and in 1924–39, declined in 1939–50 and

[13] For a fuller exposition of the measure, see Hannah and Kay, op. cit.
[14] The measure has affinities to other commonly used measures of concentration: $\alpha = 2$ is the familiar Herfindahl index and $\alpha = 1$ (or, more strictly, $\alpha \to 1$) is equivalent to the entropy index of concentration. For $\alpha = 0$ firms acquire importance simply by existing: the index then equals the number of firms.
[15] The coverage of the quoted population probably also increases because the constant samples of quoted companies are growing faster than the rest of manufacturing industry, but it is not possible to correct for this possible source of error, which would tend to understate increases in concentration and exaggerate declines in concentration.

TABLE A.3 *Concentration among quoted firms in the UK, 1907–55*

Elasticity (α)	1907	1924 constant	1924	1939 constant	1939	1950 constant	1950	1955 constant
0	571	456	726	516	1712	1502	2103	1981
0·6	287	156	274	173	542	599	871	663
0·8	224	112	198	121	354	423	612	440
1·0	174	82	146	87	232	297	422	292
1·2	136	63	110	64	156	210	292	200
1·4	107	50	85	48	110	154	207	145
1·6	86	41	68	38	81	117	153	111
1·8	70	35	56	31	63	93	119	89
2·0	58	30	47	26	51	77	96	75
2·5	40	23	34	19	35	54	66	55

Note: The lower the index, the *higher* the concentration.
Source: See text.

began to rise again in 1950-5.[16] These results are consistent with the direction of change between the turning points indicated in the simpler measure of concentration employed in Table A.2. Since the sample of quoted firms, on which Table A.3 is based, accounts for a large portion of manufacturing industry (perhaps as much as half in the later years), the consonance of the results is reassuring.

Unfortunately the years for which size distributions of firms have been computed in this way do not give a full measure of the trend in those two decades – the 1920s and 1960s – in which, according to Table A.2, concentration was increasing most rapidly. This is, in the first place, because there is no estimate for any turning point between 1924 and 1939 (a period which probably includes an initial rise, followed by a fall in concentration); and, in the second place, because studies of later years (which share the major drawback of the earlier study by Hart and Prais) have not been reconstructed using a more appropriate index of concentration.[17] However, drawing on a recent study of the periods 1919-30 and 1957-69 by J. A. Kay and the present author,[18] we can complete the picture for the years in which concentration increased most rapidly. The measures of the sizes of firms used for 1919-30 and 1957-69 were, respectively, market valuations and net assets. The population in both periods consists of the larger, principally quoted, companies. Unlike Table A.3, however, it excludes non-manufacturing companies, and also the populations in each pairwise comparison are an approximately constant sample of companies so that the entire populations at each date in this case can be compared directly.[19]

16 Similar trends were originally reported by Hart and Prais ('The measurement of business concentration') so that, in this case at least, the failings of the variance of logarithms as a measure of concentration were not overwhelming.

17 Among the studies of later periods emulating Hart and Prais in using the variance of the logarithms of firms' sizes as a measure of concentration are: M. A. Utton, 'The effects of mergers on concentration: UK manufacturing industry, 1954-1965', *Journal of Industrial Economics*, vol. 20 (1971); J. M. Samuels and A. D. Chesher, 'Growth survival and size of companies 1960-9', in K. Cowling (ed.), *Market Structure and Corporate Behaviour* (1972).

18 Hannah and Kay, *Concentration in Modern Industry*. The findings of this study on 1957-69 were broadly consistent with the findings of the studies mentioned in note 17. Where they were not so consistent, the discrepancy could be accounted for by inappropriate grouping of data or misleading concentration indicators used in these earlier studies.

19 This applies to comparisons between 1919 and 1930 and between 1957 and 1969. However, as the coverage of the sample in the earlier period is less than in the later period, it would be misleading to make a comparison between, say, 1930 and 1957 on the basis of this data.

In Table A.4, by comparing columns 3 and 4, we can see a substantial increase in concentration for all values of α within the relevant range, and the impression of a strong upward trend in concentration between 1919 and 1930 and between 1957 and 1969 (suggested earlier in Table A.2) is thus confirmed.

Table A.4 also shows estimates of the contribution of merger activity to increases in concentration. In Chapters 7 and 10, using the simple (but arbitrary) concentration ratio, we suggested that in both the prewar

TABLE A.4   *Changes in concentration in a sample of UK manufacturing firms, 1919–69*

1 Period	2 Elasticity (α)	3 1919 population	4 1930 population	5 Contribution of mergers
1919–30	0	1263	584	−671
	0·6	645	247	−350
	0·8	506	182	−274
	1·0	395	135	−212
	1·2	312	101	−167
	1·4	249	78	−131
	1·6	203	63	−105
	1·8	169	52	−86
	2·0	144	44	−73
	2·5	104	33	−50
		1957 population	1969 population	Contribution of mergers
1957–69	0	1182	744	−439
	0·6	580	326	−268
	0·8	436	245	−206
	1·0	324	187	−152
	1·2	241	146	−108
	1·4	182	118	−76
	1·6	141	97	−53
	1·8	112	82	−38
	2·0	92	71	−22
	2·5	63	52	−14

Note: The lower the index, the *higher* the concentration.
Source: Hannah and Kay, *Concentration in Modern Industry.*

and postwar periods mergers were the most significant cause of rapidly increasing concentration. It is also possible, using Table A.4, to see the impact of mergers on the more comprehensive index of concentration which we have proposed. In both periods, mergers appear as the most important cause of increasing concentration[20]: at $\alpha = 1$ in 1919–30, for example, the index falls from 395 to 135 and 212 of this fall is due to merger. In 1957–69, for all values of α, the contribution of merger actually exceeds the fall in the index, implying that concentration would have declined in the absence of merger.

All of the measures of concentration which we have outlined thus far use aggregated populations, usually of a large sector of manufacturing industry or manufacturing as a whole. However, it might reasonably be argued that the economic relevance of concentration within such a population is limited, particularly in relation to concepts of competitive behaviour and monopoly power.[21] The degree of market competition between a manufacturer of cotton textiles and one of warships is, for example, unlikely to be significant, yet both of them appear together in our manufacturing industry classification. Ideally, then, we would disaggregate the population into groups of products with similar cross-elasticities of demand; but in practice the studies of individual industries which have been published have been based on the official census classification of industries (which is determined more by the technical characteristics of the product than by cross-elasticities of demand). We have already referred extensively to the scholarly work on changes in concentration disaggregated to three-digit industry level on the basis of data available from the *Census of Production* in the period since 1935. Hitherto, however, no estimates of changes in concentration disaggregated to individual industry level have been available for the period before 1935.[22] Table A.5 goes some way to remedying this omission by disaggregating the size distribution of firms for 1919 and 1930 into

[20] Only a portion of a merger activity could be included in each period, and, although most of the effects of the large mergers are undoubtedly captured in the table, many small and medium-sized mergers have been omitted. The percentage contribution of mergers shown in the table is thus a lower bound estimate.

[21] There are, however, difficulties in equating concentration with monopoly – see pp. 186–92 above.

[22] There have, of course, been studies of concentration before 1935 in single industries, for example by G. J. Stigler ('The economic effects of the antitrust laws', *Journal of Law and Economics*, vol. 9 (1966), as reprinted in his *Organisation of Industry* (Homewood, Illinois, 1968), pp. 271–95), but none for the whole range of manufacturing industries.

fifteen broadly defined industrial groups. The population of firms is the same as that used in the computations presented in Table A.4, using market valuations as a measure of size and including only the larger, principally quoted, firms.[23] Table A.5 corroborates the view that the industrial movements of the 1920s were widespread. Concentration (measured in the table by five-firm concentration ratios) increased in all of the fifteen industrial groups (columns 2 and 3). Comparison with the counterfactual population separating out the impact of mergers (column 4)[24] suggests that in twelve of them merger was the dominant cause of increasing concentration (column 5). These results are broadly confirmed by the more comprehensive measure of concentration,

$$K = \left(\sum_i s_i^\alpha \right)^{\frac{1}{1-\alpha}}$$

The only exception is that in this case for the lower values of α in the textile industry mergers appear as the most significant cause of increasing concentration, whereas in Table A.5 (and for higher values of α) they appear as only a minor contributor.[25] This contrast reflects the fact that, while the largest firms in textiles, such as J. & P. Coats and Courtaulds, relied on internal growth as a source of expansion, mergers were of greater importance to the medium-sized firms, such as the Lancashire Cotton Corporation, which are emphasized at low values of α. Hence in twelve of the fifteen industry groups (or thirteen depending on the view one takes about the elasticity α in the case of textiles), it is clear that the merger wave of the 1920s had a substantial impact on concentration.

[23] Because of differences in the coverage of the sample in different industries, it would be unwise to use the data in Table A.5 to compare the level of concentration *between* industries in 1919 or 1930, but they can safely be used to measure changes *within* individual industries (strictly, among larger firms *within* industries) over time.

[24] The counterfactual 'merged' population was constructed in a manner similar to that described on pp. 112–13 above.

[25] Hannah and Kay, *Concentration in Modern Industry*.

TABLE A.5 *Concentration in industry groups, 1919–30*

	Shares of the largest five firms			
1 Industry	*2* 1919 population (%)	*3* 1930 population (%)	*4* Counter- factual 'merged' population (%)	*5* Proportion of the increase due to mergers (%)
Food	39·0	74·0	67·7	82
Drink	25·7	40·6	33·9	55
Tobacco	94·5	99·7	97·7	62
Chemicals	61·3	86·3	84·0	91
Metal manufacture	28·7	45·9	42·7	81
Non-electrical engineering	46·3	56·2	53·7	75
Electrical engineering	43·8	51·9	56·8	160*
Shipbuilding	64·6	89·7	84·9	81
Vehicles	34·1	66·5	41·2	22
Metal goods not else- where specified	68·0	87·1	80·8	67
Textiles	47·4	64·0	51·3	23
Clothing and footwear	33·1	58·0	40·1	28
Building materials	59·5	83·0	81·4	93
Paper and publishing	40·8	73·8	64·6	72
Miscellaneous manufacturing†	49·2	84·8	70·5	60

Source: Unpublished study by M. Ackrill, L. Hannah and J. A. Kay.
Notes:
* A merger contribution above 100 per cent implies that, but for merger
  activity, concentration would have declined in this industry.
† Includes industry groups conventionally listed separately as 'leather, leather
  goods and fur' and 'timber and furniture' in addition to the conventional
  'miscellaneous manufacturing' category.

Select bibliography

ನ

This bibliography is intended as a brief guide to the literature on the history and economics of the corporate economy. It does not attempt to list all the sources used in the writing of the present work, since this bibliographical requirement has been largely catered for in the footnote references themselves. In order to minimize the necessary search time, full references, including place and date of publication, have been included in the first mention of each book or article *in each chapter*. Readers requiring further guidance in the technical literature are referred to the standard annotated bibliography compiled by the Board of Trade, *Competition, Monopoly and Restrictive Practices, a Select Bibliography* (1970). Except where otherwise stated, the place of publication of items mentioned, both in the footnotes and in this bibliography, is London.

There are a number of texts covering the historical background to the British economy in this period. D. H. Aldcroft and H. W. Richardson, *The British Economy 1870–1939* (1969), contains a useful survey of research on modern economic growth, and reprints a number of the authors' articles on the new industries and economic expansion between the wars. The evidence of retardation in Britain's growth before 1914 is discussed in D. H. Aldcroft (ed.), *The Development of British Industry and Foreign Competition 1875–1914* (1968), and A. L. Levine, *Industrial Retardation in Britain 1880–1914* (1967); though some commentators take a more favourable view of this period. W. Ashworth, *An Economic History of England 1870–1939* (1960), for example, uses a traditional historical approach to defend the performance of the British economy before 1914, while D. N. McCloskey, 'Did Victorian Britain fail?', *Economic History Review*, vol. 23 (1970), uses techniques in the so-called 'new' economic history to the same purpose. S. Pollard's *The Development of the British Economy 1914–1950* (1962) provides a useful coverage of the later period, and D. Landes, *The Unbound Prometheus* (Cambridge, 1970), sets the whole subject in an international context. The economist C. P. Kindleberger reflects on some key historical issues in his *Economic Growth in France and Britain 1851–1950* (1964). Marxist interpretations may be found in M. Dobb, *Studies in the Development of Capitalism* (1946), and E. J. Hobsbawm, *Industry and*

Empire (1968). C. L. Mowat, *Britain between the Wars 1919–1940* (1955), and A. Marwick, *Britain in the Century of Total War 1900–1967* (1968), provide useful accounts of the political and social background.

Economists have written much on the subject of industrial structure, though their writing usually lacks a sense of historical context and much of it is policy oriented. Perhaps the best text book of industrial organization is F. M. Scherer, *Industrial Market Structure and Economic Performance* (Chicago, 1970), though the majority of his examples are American, as are those of G. J. Stigler's useful collection of essays, *The Organisation of Industry* (Homewood, Illinois, 1968). A prominent British competitor is K. D. George, *Industrial Organisation: Competition, Growth and Structural Change in Britain* (1971). A. Marshall, *Industry and Trade* (1919), and E. A. G. Robinson, *The Structure of Industry* (1931), are evergreens; and a number of empirical studies by British economists are of abiding interest for the economic historian, notably P. S. Florence, *The Logic of Industrial Organisation* (1933), and its successor *The Logic of British and American Industry* (rev. ed. 1961); and G. C. Allen, *The Structure of Industry in Britain* (2nd ed. 1966). Two interesting, but neglected, theoretical contributions to the study of industrial organization may be found in R. H. Coase, 'The nature of the firm', *Economica*, vol. 4 (1937), reprinted in American Economic Association (ed.), *Readings in Price Theory* (1953), and G. B. Richardson, 'The organisation of industry', *Economic Journal* vol. 82 (1972). R. Marris, *The Economic Theory of 'Managerial' Capitalism* (1964), is an attempt to modify traditional theories of the firm to take account of institutional changes, and this, and other revisions of economic models, are discussed in R. Marris and A. Wood (eds), *The Corporate Economy* (1971). A wider view is taken in some more speculative, and readable volumes on large British and American corporations, written from a variety of viewpoints. P. A. Baran and P. M. Sweezy, *Monopoly Capital* (Harmondsworth, 1968), is a Marxist essay on the American social and economic order, while J. K. Galbraith, *The New Industrial State* (1967), and J. A. Schumpeter, *Capitalism, Socialism and Democracy* (1943), have a meliorist outlook. G. Bannock, *The Juggernauts* (1971), provides ammunition for the critics of corporations from his experience as an economist working in industry. R. E. Caves, 'Market organization, performance and public policy', in the Brookings Report (R. E. Caves, ed.), *Britain's Economic Prospects* (1968), is a balanced view of recent trends in Britain. E. T. Penrose, *The Theory of the Growth of the Firm* (Oxford, 1959), is a pioneering attempt to study the limits to the size of the firm in a dynamic context, and G. B. Richardson, *Information and Investment* (1960), provides a suggestive treatment of the relationship between industrial structure and the efficiency of investment decisions.

On the historical front, there is nothing to match A. D. Chandler's classic description of the evolution of enterprise structure in the United States, *Strategy and Structure, Chapters in the History of Industrial Enterprise* (Cambridge, Mass., 1962); but there are a number of studies in the history of management and of individual businesses which make the darkness less than total. D. F. Channon, *The Strategy and Structure of British Enterprise* (1973), is confined to post-1950 developments; and G. Turner, *Business in Britain* (1969), also covers this period well, as do the various case studies included in R. S. Edwards and H. Townsend, *Business Enterprise* (1958), and in their companion volume *Studies in Business Organisation* (1961). For the earlier period the best sources are the many histories of individual businesses, though these differ tremendously in quality, from the adulatory public relations exercises to the serious and critical scholarly studies. Among the most important of the latter are: D. C. Coleman, *Courtaulds, An Economic and Social History* (2 vols, 1969); C. Wilson, *The History of Unilever* (2 vols, 1954), and its sequel *Unilever 1945–1965* (1968); W. J. Reader, *Imperial Chemical Industries, A History* (2 vols, 1970 and 1975); P. Mathias, *Retailing Revolution* (1967); and B. W. E. Alford, *W. D. & H. O. Wills and the Development of the UK Tobacco Industry 1786–1965* (1973). Written more obviously from the perspective of the present, but nonetheless historically illuminating for that, are G. Turner, *The Leyland Papers* (1970), and R. Jones and O. Marriott, *Anatomy of a Merger, A History of GEC, AEI and English Electric* (1970). All of these studies have much to say on the role of individual entrepreneurs in the growth of firms, a theme which is more explicitly treated in P. W. S. Andrews and E. Brunner, *The Life of Lord Nuffield* (Oxford, 1955). R. A. Church, *Kenricks in Hardware, A Family Business* (Newton Abbot, 1969), gives equally scholarly treatment to a medium-sized business and thus provides us with some insight into the kind of firm about which much is said but little is known. Studies of individual industries can also be helpful and they are almost as common as business histories. Amongst the more interesting may be listed: W. Minchinton, *The British Tinplate Industry* (Oxford, 1957); H. R. Edwards, *Competition and Monopoly in the British Soap Industry* (Oxford, 1962); G. Maxcy and A. Silberston, *The Motor Industry* (1959); J. E. Vaizey, *The*

Brewing Industry 1886–1951 (1960); and J. C. Carr and W. Taplin, *A History of the British Steel Industry* (Oxford, 1962). Various collections of papers also include useful studies of individual industries, including P. L. Cook and R. Cohen, *The Effects of Mergers* (1958), and D. L. Burn (ed.), *The Structure of British Industry* (2 vols, Cambridge, 1958).

The historical experience of merger activity is still most fully described in contemporary works. The files of the *Economist* and the *Investors Chronicle* are perhaps the most useful among periodicals. H. Macrosty, *The Trust Movement in British Industry* (1907), has a comprehensive coverage of the earlier period, while P. Fitzgerald, *Industrial Combination in England* (1927), and A. F. Lucas, *Industrial Reconstruction and the Control of Competition* (1937), are sketchier. On the more recent period, there are G. D. Newbould, *Management and Merger Activity* (Liverpool, 1970), and M. A. Utton, 'Mergers and the growth of large firms', *Bulletin of the Oxford University Institute of Statistics*, vol. 34 (1972). Other recent publications which survey the historical experience of mergers and concentration are: M. A. Utton, 'Some features of the early merger movements in British manufacturing industry', *Business History*, vol. 14 (1972); P. L. Payne, 'The emergence of the large-scale company in Great Britain, 1870–1914', *Economic History Review*, vol. 20 (1967); L. Hannah, 'Mergers in British manufacturing industry 1880–1918', *Oxford Economic Papers*, vol. 26 (1974); P. E. Hart, 'Business concentration in the United Kingdom', *Journal of the Royal Statistical Society*, series A, vol. 123 (1960); and L. Hannah and J. A. Kay, *Concentration in Modern Industry: Theory, Measurement and the UK Experience* (forthcoming 1976). S. J. Prais, 'A new look at the growth of industrial concentration', *Oxford Economic Papers*, vol. 26 (1974), surveys the intellectual history of economic models of the concentration process. The reasons for the early absence of takeover bids in Britain are discussed in L. Hannah, 'Takeover bids in Britain before 1950', *Business History*, vol. 16 (1974); while A. Singh, *Takeovers, Their Relevance to the Stock Market and the Theory of the firm* (Cambridge, 1971), provides an econometric analysis of later takeover activity. The role of capital market imperfections in inducing more rapid industrial concentration in the United States is discussed in L. Davis, 'The capital markets and industrial concentration: the US and the UK, a comparative study', *Economic History Review*, vol. 19 (1966), though this is self-confessedly only a very partial explanation of the differences in the experience of the two countries.

Other aspects of corporate behaviour continue to attract the attention both of economists and of historians. J. Jewkes, D. Sawers and R. Stillerman, *The Sources of Invention* (1958), go further than their evidence justifies to belittle the achievement of large corporations in the field of research and development; while M. Sanderson, 'Research and the firm in British industry, 1919–1939', *Science Studies*, vol. 2 (1972), perhaps overcompensates. M. J. Peck, 'Science and technology', in R. E. Caves (ed.), *Britain's Economic Prospects* (1968), is a critical review of more recent British research efforts. A. Silberston surveys the modern evidence on 'Economies of scale in theory and practice' in *Economic Journal*, vol. 82, supp. (1972), though for the historical evidence on this subject one has to turn to studies of individual businesses and particular industries. The various *Reports* of the Monopolies Commission (1950, continuing) provide a useful survey of scale economies and monopolistic practices, though more recently the Commission has curtailed its historical inquiries and focused its attention more closely on the recent practices of the firms investigated. Industrial profitability, including its relation to concentration and competition, is discussed in two works with very different approaches: P. E. Hart (ed.), *Studies in Profit, Business Saving and Investment in the United Kingdom, 1920–1962* (2 vols, 1965 and 1968); and A. Glyn and B. Sutcliffe, *British Workers, Capitalism and the Profits Squeeze* (Harmondsworth, 1972). The bare facts of government intervention in industry are chronicled in J. W. Grove, *Government and Industry in Britain* (1962), but a more interesting approach can be found in the writings of the politicians themselves, the classics being the Liberal Industrial Inquiry's report *Britain's Industrial Future* (1928) and C. A. R. Crosland's *The Future of Socialism* (1957). The Conservatives have produced no work of similar quality on the political implications of the growth of corporations, though A. Marwick, 'Middle opinion in the thirties, planning, progress and political agreement', *English Historical Review*, vol. 79 (1964), and N. Harris, *Competition and the Corporate Society: British Conservatives, the State and Industry 1945–64* (1972), suggest that some Conservatives were also developing new ideas in response to contemporary economic changes. The development of ministerial views on relations with industry can now be followed in the files of the Cabinet and of the Board of Trade at the Public Record Office, for the period up to and including the Second World War. The *Reports* of the Macmillan Committee on Finance and Industry (Cmd. 3897, 1931) and of the Balfour Committee on Industry and Trade (Cmd. 3282, 1929) should be read together with the earlier and fuller research reports of the latter committee: *Overseas Markets* (1925); *Industrial Relations* (1926); *Factors in Industrial and Commercial Efficiency* (1927); *Survey of Metal Industries* (1928); *Survey of Textile Industries* (1928); and *Further Factors in Industrial and Commercial Efficiency* (1928). International investment by corporations is a subject on which little was written until the postwar period, though J. H. Dunning, *American Investment in British Manufacturing Industry* (1958), provides a useful historical survey, and the essays collected in J. H. Dunning (ed.), *International Investment* (Harmondsworth, 1972), provide a thorough survey of more recent theoretical and empirical work on the subject.

Index

ಎಲ

Index

Renold and Coventry Chain Co. 42, 135

Renold family 42, 135n

research and development 1, 9, 40, 72, 80, 91, 95, 127–9, 138, 140–1, 153, 159, 168, 184–5, 198n, 230

restrictive practices 6, 17, 57, 59, 96, 97, 155–8, 162, 169, 175, 187, 191

Restrictive Practices Court 169, 187

retail sector 111n, 190, 191

Revell, J. 62n, 64n

Rhodesia 127

Richardson, G. B. xi, 14n, 82n, 126n, 127n, 160n, 227, 228

Richardson, H. W. 71, 115n, 116n, 122n, 226

Richard Thomas 118, 145

Rist, A. 72

road transport 111, 130, 188

Robbins, Lionel 52, 59n

Robertson, D. H. 88n

Robinson, E. A. G. 32n, 79n, 227

Robinson, J. 52n

Robson, R. 74n

Robson, T. B. 97n

Rockefellers 24

Rogers, John 92

Rolls Royce 198n

Roosevelt, F. D. 33

Rose, H. B. 17n

Rose, T. G. 96n

Rosenbluth, G. 103n

Rostas, L. 153n, 163n

Rothermere, Lord 131

Roundway, Mr 38n

Rowley, C. K. 176n

Royal Commission on Food Prices 50n, 51n

Royal Commission on the Press 131n, 149n

rubber industry 103, 118, 120

Rubinstein, H. xi

Runciman, Walter 57

Rutherford, Lord 128

Ryan, John 85, 147n

Ryland, T. H. 51n

Salt Union 27

Salter, W. E. G. 161n, 192n

Samuel, R. 198n

Samuels, J. M. 196n, 221n

Sandberg, L. 152n

Sanderson, M. 127n, 140n, 230

Saul, S. B. 42n

Sawers, D. 129n, 159n, 184n, 230

Sawyer, M. C. 165n

Sayers, R. S. 129n

Scherer, F. M. 151n, 181n, 227

Schumpeter, J. A. 197n, 227

scientific management 32–6, 37, 95, *see also* management, rationalization

Scopes, Sir Frederick 135n

Scott, J. D. 55n, 138n, 154n

Seager, Sir William 36n, 38n

Sears, J. 121

Securities Management Trust *see* Bankers' Industrial Development Co.

Select Committee on High Prices and Profits 49n

self-government of industry 155

Seltzer, L. H. 66n

Shannon, H. A. 18n

share prices 22, 66–9, 117n, 170, 206

Sheffield Smelting Co. 147n

Shell 117n

Shepherd, W. G. 103n

Sherman Act 46

shipbuilding industry 44, 69, 118–20, 138, 148, 172, 225

Shonfield, A. 164n

Siemens Dynamo Works 125

Silberston, Z. A. xi, 27n, 71n, 122n, 139n, 140n, 159n, 161n, 167n, 177n, 228, 230

Silverman, R. 188n

Simonson, P. F. 54n

Singh, A. 106n, 165n, 181, 229

Slater, Jim 171

Sloan, A. 96n

small firms 1, 8, 13, 36, 146, 153, 166, 168, *see also* family firms

Smith, Adam 11, 19–20

Smith, Vivian 157–8

Smyth, D. J. 196n

Snowden, Philip 53, 55

soap industry 17, 45–6, 48, 83–4, 103, 117n, 130, 228

Social Science Research Council xi

socialism 39, 51, 199, *see also* Labour, Marxists

Solvay process 27

SPD Co. 130

Sperling & Co. 40, 69

Stacey, N. A. H. 89n, 90n

Index

Vice, A. 170n

Vickers 25, 30, 44, 72, 89, 118, 120, 121, 126, 137, 154n

Vickers-Armstrong 55n, 138, 172

wages 36–7, 45, 46

Walker-Cain 120

Walker, G. 97n

Walker, J. L. 204n

Wallpaper Manufacturers 17, 23, 121, 154

Walshe, G. 165n, 166n, 192n

war 29–31, 47, 152, 156–7, 173

Ward, Harry xi, 34n, 40n, 56n, 82n, 87n

Warriner, D. 41n

Watney Combe Reid 25, 26, 118, 120

Watson, A. 32n

wealth *see* concentration of wealth

Webb, Sidney 48

Weber, A. F. 16n

Wedgwood, Josiah 89n

weighing machines 103, 117n

Westinghouse 125, 132

Whitbreads brewery 8

Whitehead, H. 87n

Whittington, G. 111n, 165n, 184n

Whyte, A. G. 55

Wiles, P. J. D. 156n

Williams, B. R. 158n

Williams, H. G. 50n

Williams, P. M. xi, 46n

Williamson, O. E. 85n

Wills, W. D. & H. O. 15–16, 42n, 61, 228

Wilson, C. 4, 45n, 76n, 90n, 97n, 130n, 139n, 154n, 228

Wilson, Sir Horace xi, 56, 57

Wilson, R. 82n

Wilson, R. E. 147n

Witton 125

Wolseley 138

Wood, A. 85n, 227

workers 2, 31, 81, 198

Worswick, G. D. N. 216n

Worthington 129

Wright, A. 31

Wright, J. F. xi, 170n

Wright, P. D. xi, 133

Yamey, B. S. 46n, 169n

Yorkshire Copper Works 209

Yorkshire Imperial Metals 209

Yorkshire Woolcombers' Association 23

Young, Allyn 14n

Young, A. P. xi

Zinkin, M. 12n